PHYSICS: A SHORT HISTORY FROM
QUINTESSENCE TO QUARKS

Physics
A Short History

FROM QUINTESSENCE
TO QUARKS

J. L. HEILBRON

OXFORD
UNIVERSITY PRESS

OXFORD
UNIVERSITY PRESS

Great Clarendon Street, Oxford, OX2 6DP,
United Kingdom

Oxford University Press is a department of the University of Oxford.
It furthers the University's objective of excellence in research, scholarship,
and education by publishing worldwide. Oxford is a registered trade mark of
Oxford University Press in the UK and in certain other countries

First Edition published in 2015

Impression: 1

Published in the United States of America by Oxford University Press
198 Madison Avenue, New York, NY 10016, United States of America

British Library Cataloguing in Publication Data
Data available

Library of Congress Control Number: 2015939044

ISBN 978-0-19-874685-0

Printed in Great Britain by
Clays Ltd, St Ives plc

CONTENTS

CONTENTS

LIST OF FIGURES

The Scale's but small, Expect not truth in all.

Wenceslaus Hollar, *A new map of the Cittiess of London and Westminster*, 1676

Precision is not to be sought for alike in all discussions.

Aristotle, *Nichomachean ethics*, 1094b12

INTRODUCTION:
THE GREEK WAY

————•◦•————

The Superconducting Super Collider (SSC), the dream of American high-energy physicists, would have had a circumference of 87 kilometers and a price tag of billions of dollars. Its proponents justified the expenditure on several grounds. On the high ground, it would probe the universe to philosophical depths and thus "keep faith with the Greeks." Below ground, it would advance tunneling technique and give society perfect sewers. Congress cancelled it in 1993.

The undertaking represented by the SSC might be the way to an ultimate physics. But it is not the Greek way. In antiquity, physics was philosophy, a liberal art, the pursuit of a free man wealthy enough to do what he wished. He did not aim to improve sewers and, since he had no need of public money, did not have to claim that he would. Nor did he want apparatus, since he seldom experimented, or mathematics, since he seldom calculated. The few ancient applications of mathematics to physics constituted a mixed science devoted to the description of phenomena rather than to the search for principles.

In the tripartite division of Greek philosophy, physics stood between logic and ethics. It inquired into the principles regulating the physical world from the high heavens to the Earth's center, and from the human soul to the life of the least of living creatures. It thus functioned as a natural theology—defining man's place in nature—and as a necessary approach to ethics, or the principles of a good life. For two millennia the main practical value of physics lay in the ethical consequences of its versions of the way the world began and persists.

Greek physics, with its eye to ethics, its indifference to mathematics and experiment, and its independence of states and courts, is sufficiently distinct from an enterprise conducted by salaried teams requiring elaborate technologies and mathematical analyses to deserve a different name. Let it be *physica* and those who cultivated it *physici*. This short book describes some of the ways by which ancient *physica* became modern physics. It does not ransack history to find items in ancient and medieval science that look like physics, but sketches the place and purpose of *physica* in the societies that supported it. Hence the primary site(s) of cultivation receive special emphasis: the independent private school (antiquity), court and library (Islam), university (later Middle Ages), court again (Renaissance), academy (late seventeenth and eighteenth centuries), university again (modernity), and university-government-industry (postmodernity). Of course, successive forms did not annihilate their predecessors. Academies of science survive, primarily as honorific societies and depositories of history, although a few flourish as national channels for funding, consultation, and outreach. The scientific advisory apparatus of government may be considered the descendant of courtly science; and the Greek schools, with their characteristic discursive style, continue in the myriad seminars in which the world's nascent science is presented and anatomized.

The story of the SSC may do service as a double symbol: through its cost, size, and design, of the transformation of *physica* into physics; through its cancellation, of the ongoing dilution of the dominance of the United States within an increasingly competitive international system.

Fig. 1. School of Athens. Raphael's evocation of the intellectual vigor, diversity, and discursiveness of Greek Science.

1

INVENTION IN
ANTIQUITY

———•◦•———

Tradition follows Aristotle in identifying the earliest *physici* as some gentlemen of Miletus, a flourishing Greek city on the coast of Anatolia, and in specifying half a dozen other Greek speakers as their successors. In this philo-Hellenic creation myth, no Greek *physicus* learned anything of any importance from a barbarian during the 250 years between the times of the eldest Milesian, Thales, and Aristotle. The story that Pythagoras, if he existed, did so partly in Egypt, suggests outside input; and studies of cuneiform texts reveal a natural knowledge among the Babylonians in some ways more advanced than that of the ancient Greeks. Still, the essential criterion that Aristotle used to identify his predecessors was not that they were Greek, but that they had conquered a paralyzing prejudice. Despite robust contrary evidence, they believed that the natural world runs on law-like principles discoverable by the human mind and immune from interruption or cancellation by meddling gods and demons.

Whether or not the Milesians deserve the blame or credit for this bold departure, it underlies and circumscribes all forms of *physica*, natural philosophy, and physics. Its implications go farther even than replacing caprice by law-like behavior. Since the gods displayed all too faithfully the traits and behavior of human beings, de-deifying implied (to speak Greek) deanthropomorphizing. The progress of physics has continued to remove human quirks and qualities projected on to nature. Thus nature, or the objective world, came to lose not only benevolence, malevolence, and color, but also such apparently indispensable attributes as space, time, and causality.

Of the four main schools of ancient philosophy, Aristotle's paid greatest attention to *physica* (see Figure 1). Having a particular interest in zoology, he derived his fundamental principles with an eye to the classification of animals. Because of his emphasis on *physica* and because his philosophy dominated during the Middle Ages and beyond, convenience advises taking it as normative. In antiquity, however, it had to compete with Platonic, Epicurean, and Stoic philosophies. The fluctuating market share of a school reflected the reputation of its leader as well as fad and fortune. Typically, its premises, library, and other assets passed from the founder to his senior disciples, and, perhaps consequently, the schools bore names suggestive more of real estate than of scholarship: the Academy (Grove) for the Platonists, Lyceum (a shrine) and Parapatos (a place for walking) for the Aristotelians, the Stoa (Porch) for the Stoics, and the Garden of Epicurus.

During the 800 years from the founding of the Academy in Athens and its refounding in Alexandria, the four schools underwent many reversals of fortune, including dissolutions and amalgamations. When operating, they were forums for free discussion of the sort practiced in political circles, but centered on studies pursued for personal improvement rather than for civic or financial advancement. Although they had members who studied with them for decades, they also admitted students who sampled each in turn. When Romans like Cicero frequented the schools of Athens they took advantage of this *Lernfreiheit*. The Pythagoreans did not form such a school, as they did not tolerate deviations from their doctrines and way of life, and acted ruthlessly against members who revealed their secrets.

Physica

Although *physica* ran from astronomy through zoology to psychology, limitation of coverage to cosmology and cosmogony, as in this book, is not unacceptably anachronistic provided we recognize that the same principles of structure and change applied to all natural processes. This truncated *physica* corresponds to the books of Aristotle dealing with general principles (*Physica*), the heavens (*De caelo*), the region between the Moon and the Earth (*Meteorologica*), and the creation and destruction of things on or in the Earth (*De generatione et corruptione*). These books (and the rest of the corpus from logic to ethics) became available in a standard format edited around 60 BCE from the lecture notes Aristotle bequeathed to his successors. They constitute the main parts of a theory of everything, or, as the moderns say, a TOE.

Cosmologies

The first TOEs Aristotle stepped on belonged to the Milesian monists. He then took on more generous materialists: Leucippus and Democritus, who allowed two principles, atoms and the void; Empedocles, who accepted the three Milesian elements (water, air, fire) and added earth to complete the tetrad; and Anaxagoras, who admitted an infinite number of different sorts of stuff. There were also those whose matter had no stuff at all, Pythagoras notably, for whom number had an independent existence—a concept that made no sense to Aristotle. The Pythagoreans' deduction that there must be a counter earth circulating opposite ours around a central fire (to raise the number of heavenly bodies to the holy tetractys) proved to Aristotle both the falsity of their *physica* and the nonsense to which numbers can lead.

Aristotle's teacher, Plato, had some sympathy for Pythagorean numerology and a belief in the importance of mathematical patterns for cosmological architecture. He took as his material not only mathematical abstractions but also supersensible idealizations of classes of objects: for example, the Idea "Horse," in which individual horses "participate" more or less, but always imperfectly. Consequently, although Plato was optimistic about the possibility of knowing the ideal world and even the supreme Good that made the Ideas and their relations intelligible, he did not allow the possibility of true knowledge of the things of this world. Since no material individual could express an Idea perfectly, our *physica* can never be other than fuzzy.

Physica comes from "physis" meaning "nature," which, according to Aristotle, "is the source or cause of being moved or being at rest." What makes things move? The early *physici* adumbrated four causes of change that Aristotle later codified. The monists and the atomists considered only the material cause. Empedocles and Anaxagoras provided action by taking some principles to be active and others passive—vague glimpses of efficient causes. Others saw the need to explain order in a universe of change and hinted at a teleological or final cause, such as set by a cosmic Mind. And Plato supplied a fourth cause, the formal, the Idea in which a thing participates.

Aristotle's inventory of the cosmological ideas of his predecessors, including anticipations of the four causes of change, was not an idle retrospective. It confirmed, and even proved, that he had not overlooked anything fundamental. "Of all who have discussed principles and causes none has spoken of any kind except those which have been distinguished in [my] discourses on Physics. They are all unmistakably, though obscurely, trying to formulate these." Aristotle's TOE, thus established as complete, makes use of some special concepts. A *substance* is any individual thing. The collection of its properties constitutes its *form*, which, contrary to a Platonic Idea, occurs only in union with *matter*. Form can be divided, although only mentally, into *essence*, which makes a substance the sort of thing it is, and *accidents*, which can change without causing the substance to alter its essence or kind. The essences of the four elements are easily stated: fire is dry, hot, and absolutely light; air is hot, moist, and relatively light; water is cold, moist, and relatively heavy; earth is cold, dry, and absolutely heavy. The elements can transform into one another, as fire evaporates water into air. An essence is usually compatible with a wide range of accidents: Socrates may be warm or cold, but heated or cooled too much he will cease to be Socrates.

It may now be intelligible to state that a substance's matter and essence are its material and formal causes, and that the active qualities hotness and moistness are the principal efficient causes of change. The final cause is the purpose for the existence of a form. Heavy bodies have gravity so as to fall toward the center of the world, and light bodies have levity so as to be able to rise toward the heavens, to restore order disrupted by the activity of animate creatures or the revolutions of the celestial spheres. These spheres and the stars and planets they carry cannot be made of the four elements, whose forms require that they move when unimpeded in a straight line toward or away from the world's center. There must therefore be a fifth element, a *quintessence*, which, obeying its formal and final causes, circulates around the universal center. With these few principles and some ad hoc adjustments, Aristotle worked his way from the Mind of the Unmoved Mover, which, as it can think only of the most sublime thing, can think only of itself, down to the mind of man, which, though no less selfish than the Mind of the Universe, is subject to frequent change like everything else below the quintessential heavens. And where there is change, there cannot be certainty; the best a *physicus* can do is to find a "rule [that] applies to what is always true or true for the most part."

A few deductions from Aristotle's approximate *physica* that came under sustained scrutiny will give some impression of its general character. Every motion, whether change of place or color or species, requires an external mover. Since none is available in a vacuum, where there literally is no place (no reference material) by which a body could orient itself, there can be no vacuum. Hence the flight of an arrow implies a vortex in the air, which cedes a place to the tip while slipping in behind to provide a push at the tail. The ambiguous role of the air, offering both resistance and propulsion, made an obvious difficulty. Another awkwardness arose from the absolute dichotomy between terrestrial and celestial physics. Because the heavens cannot change, transient phenomena that appear to take place there, like comets and meteors, must have their seat with lightning and the weather in sublunar regions.

Nothing, however, is more obvious than that the Sun influences the weather. How? Sometimes Aristotle wrote as if he thought that the Sun was hot, which would violate his proscription against terrestrial qualities in heaven. More often, he ascribed the seasonal powers of the Sun to its annual revolution, which, together with the rolling of the quintessential spheres, continually stirs up the sublunary regions. These disturbances cause moist and dry vapors to rise from the earth. Precipitation results from the moist exhalation, winds from the dry. "The same stuff is wind on the earth, and earthquakes under it, and in the clouds thunder." Lightning and thunder are dry exhalations breaking free from the clouds like pips squeezed from a fruit. The rainbow is a reflection from the clouds. Aristotle described it, in a manner unusual for him, geometrically: the Sun, the eye of the observer, and the center of the bow lie on a straight line that cannot exceed a certain angle with the horizon. This factoid would have a long and influential history.

Despite the continuing operation of final causes, the world is not in perfect order. The frictional drag of the turning lunar sphere on the stationary region of fire below it produces such anomalies as fiery meteors in the air, mountains above sea level, and water below earth. In the big picture, however, the universe resembles an onion. Peeled from the outside in, it discloses the fixed stars, the planets and luminaries in the conventional order Saturn, Jupiter, Mars, Sun, and Moon, and, in slight disorder, the elements Fire, Air, Water, and Earth. What is outside the skin? Here the onion analogy fails. There is literally nothing there. And just as there is no space not included in the visible universe, there was no time at which it did not exist.

Cosmogonies

Aristotle's world picture thus lacked a creator. So did the atomic theory of Democritus, who, nevertheless, offered a creation story for the visible universe. It began to be when a clutch of the infinity of atoms bouncing about in the infinite void happened to come together in a great vortex from which the largest fell to the center, forming the Earth. From what remained, centrifugation produced the air, the luminaries, the planets, and the stars. Although all things are made of the same dull stuff, they appear different to us because our sensory systems can build rich images from the few properties (size, shape, motion) of their constituent atoms. Epicurus added a spontaneous "swerve" to explain how atoms, falling in parallel through the void, occasionally collide and concatenate a world. According to him, the soul can exploit the swerve to choose to live the good life in an otherwise pointless universe. Since the Epicurean did not have to fear gods in this life or anything in the next, he could take moderate enjoyment of the flesh and free employment of the mind as the greatest goods. The inevitable erosion of all sound doctrine has transformed Epicurus' sober happiness into selfish hedonism, and his name into an eponym for refined pleasure.

Whereas the atomists allowed for the creation of many worlds in space and time by random accretion of their parts, the Stoics supposed that the single cosmos they admitted, geocentric like Aristotle's, is alternately destroyed and recreated. In place of disparate atoms, the Stoics put a continuous prime matter; and in place of bumps and grinds, "pneuma," a self-moving elastic compound of fire and air that gives matter its cohesive and other properties. Strict causality applies everywhere, guaranteed and effected by the spatial continuity of the pneuma, whose changing tension drives the world through repeated identical thermodynamic cycles. The system would seem to rule out free will decisively. But since ethics required a free acceptance of fate and the mental preparation necessary to meet it, Stoics had to find a way around the strict causality of their physics. Their solution was no more plausible than the Epicurean swerve. As an alterative to strict atomism, however, the Stoic concept of a space-filling active elastic spirit had a future.

In contrast to the Peripatetics with their unchanging uncreated cosmos, and the atomists and Stoics with their random and cyclical worlds, the Academics had a full cosmogony, with a creator as well as a creation story. As told by Plato's mouthpiece, the mathematician Timaeus of Locris, the Demiurge who made the realm of Ideas and the World Soul used what he had left after rolling out the celestial equator and the paths of the planets to manufacture a swarm of human souls. These he sent to stars in the realm of Ideas to await planting in bodies created by the lesser gods to whom he assigned the task of making the sensible world. At the end of life, the rational soul returns to its domicile star if its human possessor had lived a good life; if not, the soul reincarnates in a lesser being. Our animal parts, the gift of the gods, serve merely to keep our head, the seat of our reason, from rolling around on the ground.

Used properly, our rational soul can bring us through observation of the motions of the heavenly bodies to the discovery of number, time, and harmony, and to the contemplation of the Ideas. We might then perceive that the Ideas of the four elements and the quintessence are linked to the mathematics of the five regular solids. The plane faces of three of them (the tetra, octa, and icosahedron) are equilateral triangles, and consequently the elements corresponding to them (fire, air, and water) are interconvertible (see Figure 2). The remaining two, the cube and the dodecahedron, are the "Ideas" of the earth and the universe as a whole. We should not press the obvious difficulties. The lesser gods who created the material world were not entirely competent. "[I]t is fitting that we should, in these matters, accept the likely story and look for nothing further."

Roman worlds

Although foreign students from Rome added little of note to Greek *physica*, they turned much of it into useful summaries and

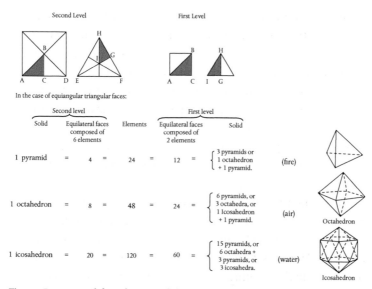

Fig. 2. Geometrical foundations of the world. The faces of the cube come in aggregates of elementary isosceles right triangles ACB; those of the pyramid, octahedron, and icosahedron come in aggregates of various sizes of elementary 30-60-90 degree triangles IGH. The diagram shows how the second level aggregates from the first and, in the case of the three solids with equilateral triangles as faces, how the solids can transform into one another. Adapted from Cornford, *Plato's cosmology* (Indianapolis: Bobbs-Merrill, nd), pp.230–9.

compendia. During the last years of the Republic, Lucretius versified Epicurus' atomism. During the reign of Caesar Augustus, the poet Ovid made a conspicuous place for Pythagoreanism in his *Metamorphoses*. Under the Emperor Nero, Seneca composed a meteorology on Stoic principles—cut short, unfortunately, by Nero's invitation (which Seneca could scarcely refuse) to commit suicide. And in the years before his fatal inspection of Vesuvius during its eruption of 79 CE, Pliny the Elder crammed into one of the 37 books of his *Natural history* a qualitative survey of the world featuring a God unpolluted by commerce with human beings, a naturalistic account of meteors, comets, and eclipses calculated to free humankind from fear of lesser gods, some hints at atomism and stoicism (*physica* of chance and necessity), and a few bars from Pythagorean music of the spheres.

Lucretius begins his poem with the customary invocation of a muse, in this case Venus, after which he disobligingly announces that his great purpose is to remove her and all the other gods from human concerns. The atoms cavorting in the infinite void follow the laws of necessity apart from the occasional unintelligible swerve that gives spontaneity to the world and freedom to the will. No external agency can interfere with this process. Therefore, although you are nothing but chance congeries of buzzing particles, be joyful, submit to fate, and fear not the impotent gods.

The lengthy exposition of Pythagorean doctrine that Ovid jams into the last book of his catalogue of beings that transform into beasts treats only briefly the founder's *physica*, his teachings about "The great world's origin, the cause of things | what nature is, what god, and whence the snow | what makes the lightning, whether thunder comes | from Jove or from the winds when clouds burst wide | why the earth quakes, what ordnance controls | the courses of the stars | and the whole sum | of nature's secrets." Most of Ovid's account of Pythagorean thought concerns metempsychosis and vegetarianism, a belief and a practice indissolubly connected. If souls transmigrate, how can you tell whether your ox, your faithful brother at the plow, was not in fact your late brother?

This sound teaching fell on deaf ears. Ovid admitted as much; Seneca, writing half a century later, reported that in his time would-be Pythagoreans could not even find a teacher. But that, according to Seneca, merely mirrored the sad state to which philosophy in general had fallen in his time: "Many philosophical lineages are dying out without a successor." His *Natural questions*, an attempt to reinvigorate Stoic *physica*, reconstructed the meteorology of the Stoa and the Lyceum to answer the questions whose solution Ovid credited to Pythagoras—the causes of the wind and weather, thunder and lightning, and earthquakes. All of these problems had been standard in *physica* since the time of Thales. They were, as they still are, of greater importance to humankind than the doings of Demiurges. In Seneca's meteorology, high winds, thunder, and earthquakes accompany the expansion of previously constrained pneuma. Breaking from Stoic authority, he enrolled comets, famous for bad reputations, among the planets and toyed with the possibility that their daily motions, and those of the stars in general, arise from a revolution of the Earth rather than from the turning of the great celestial vault. As his placement of comets suggests, the interconnectedness of the Stoics' space-filling active pneuma destroyed the barrier between the heavens and the Earth characteristic of Aristotelian cosmology.

Seneca puffed up his meteorology with moral reflections that appear to be its reason and result. If you do not inquire into the material foundation of the world and the nature of its creator or guardian, or speculate whether he still creates (if he ever did) or has retired, whether he thinks only of himself and whether he can amend fate, you enslave yourself to merely human affairs.

Life would not be worth its pain and suffering were it not for the opportunity to learn that nature, fate, the world, providence, and God are different names for the same thing; that God did not make the world exclusively for human beings; that life is a sentence of death; and that philosophy, by dispelling fear of death, is the only healthy way to resign oneself to fate. Seneca's eloquent combination of moralizing and meteorology made the *Natural questions* the authority on the physical problems it treated for most of the Latin Middle Ages.

Had Plutarch's dialogue on the markings on the Moon not suffered an eclipse of 15 centuries, Seneca's stoicism would have faced a sprightly challenge in medieval times. Plutarch freely criticized most ancient systems. As a probe and goad he asked the obvious question: why do we not see the Sun's image on the Moon as we see it on the sea? Since the answer depends on knowledge of the Moon's makeup, it was beyond the competence of mathematicians. What then did the *physici* have to say? Plutarch's spokesman replied that the Peripatetic model of the Moon as pure quintessence is nonsense; the mottled, imperfect lunar surface betrays some admixture in its makeup. The Stoic view—that it is made of a sort of *pneuma*—is worse, since air and fire cannot reflect visual rays. We are left with the Platonic account that the Moon is more "earthy" than heavenly. Its rugged surface acts as if composed of a great many randomly oriented mirrors, which can unite visual and solar rays without producing an image.

What keeps the rocky Moon from falling on the Earth? "[T]he rapidity of its revolution, just as missiles placed in slings are kept from falling by being whirled around in a circle." What keeps it close to Earth? A literal law of nature: "the position of earth lays an action against the Moon and she is legally assignable by right of propinquity and kinship to the earth's real and personal property." Is the Moon inhabited? Very likely, but by creatures as different from us as we are from fish, and also by disembodied souls improving themselves for their next incarnation. Plutarch's ideas about the Moon, minus its role as host to rehabilitating souls, recur in Galileo, Kepler, and many lesser heroes of the Scientific Revolution.

During the philosophical downturn mentioned by Seneca, the Platonic Academy did not exist. It had come to a temporary end in the destruction that accompanied the Roman conquest of Athens in 86 BCE after a period of intense skepticism about the possibility of achieving any secure knowledge about anything. It revived in 410 CE with a curriculum based on "Neoplatonism," invented in Rome by Plotinus, who died in 270 CE, and systematized by his disciple Porphyry. Plotinus was not a skeptic. His confident speculations rose above even the Demiurge, whose employment as creator, even if only of the rational realm and the lesser gods, seemed to him incompatible with occupying the acme of the divine pyramid. How did the Demiurge relate to the impassable God or Good that Plato had put in charge of everything? Plotinus' answer was that God stands immovable at the head of a chain of created and creating "emanations." The first of these emanations, Intellect, contains the Ideas; the second, Soul, contains Nature; the next is the Demiurge. Plato had not supplied much information about the natural world created by the Demiurge's collaborators, and his thesis that we cannot know such things in principle did not encourage elaboration of

Timaeus' hints about polyhedra. So Neoplatonism grafted on Aristotle's *physica* as its account of the visible universe created by the demigods employed by the Demiurge.

Another easy conflation assimilated the top of the Neoplatonic power chain—the One, the Intellect, and the World Soul—with the Christian Trinity. In the orthodox view, God did not employ a vicar in his creative works. But several Christian sects still competitive in the fourth century taught that God the Father created only spiritual beings, one of whom, Jehovah by name, broke the chain of spiritual emanations and materialized his successors. Neoplatonism was the leading philosophy, and its co-inventor Porphyry the most effective critic of Christianity, of the late empire. His opponents, the architects of Christianity concerned to achieve standardization of belief, faced the challenges of harmonizing discrepancies among the four canonical gospels, explaining evil in the creation of a beneficent deity, and defining the relationships among the Persons of the Trinity. The most abstruse of their conundrums probably could not have been concocted without the language and concepts of Aristotelian *physica* and Neoplatonic philosophy.

Applications

Physici and their speculations had less to offer to princes than applied mathematicians who could give practical advice. The greatest of them all, Archimedes, worked for a tyrant of Syracuse. Although best known now for his detection of a counterfeit crown and his legendary feats in defense of Syracuse, Archimedes preferred to be remembered in the manner of a *physicus*, as a liberal artist. According to Plutarch, he regarded his practical accomplishments as "mere accessories of geometry practiced for amusement" and "every act that ministers to the needs of life as ignoble and vulgar." Plutarch added that Plato had chastised two clever mathematicians, Eudoxus and Archytas, for turning their hands to practical things, and becoming "corruptors and destroyers of the pure excellence of geometry."

In contrast to the many useful applications of geometry, applied *physica* could only veneer a practical man. Vitruvius recommends in his *Ten books of architecture*, which dates from the early Roman Empire, that the student learn *physica* to be able to judge excellence, to achieve authority, and to grow "courteous, just, and honest." More obviously, knowledge of "the principles of physics [as taught] in philosophy" was needed for constructing waterworks; of the principles of musical intervals, for tuning catapults; of astronomy, for choosing sites for buildings; of meteorology, for avoiding places exposed to winds, lightning, and earthquakes. None of this related to the main business of construction, and for good reason. *Physica* knew nothing about the strength of materials. When it came to foundations, Vitruvius could only advise to "dig down . . . as deep as the magnitude of the proposed work seems to require."

A more lasting application of physical principles occurs in the writings of Ptolemy, who seems to have been associated with the Library and Museum of Alexandria, in effect a state institution. He distinguished his astronomy, enshrined in the 13 books of his *Syntaxis mathematica* or *Almagest* (as it is usually named after its Arabic translation), from his four books of astrological interpretation, the *Tetrabiblos*. The *Almagest* considers planets only as

moving points; the *Tetrabiblos* ascribes to them the active elemental qualities that bring their influence down to Earth. The Sun
is primarily hot and secondarily dry, the Moon primarily moist
and moderately warm, and the planets warm, moist, cold, and
dry in different degrees depending upon their distances from the
luminaries and the Earth. Judged by temperature and humidity,
Jupiter, Venus, and Moon are beneficent, Mars and Saturn evil-
doers, and Sun and Mercury ambiguous. From this astrophysics
Ptolemy derived a physical anthropology that explained why
Ethiopians are black, Scythians white, inhabitants of the temperate region medium in color, civilized, and sagacious, and
himself mathematical.

Although the astrophysics of the *Tetrabiblos* admitted earthly
qualities into the celestial regions, Ptolemy's astronomy revolved
in the quintessence of the Aristotelian universe. According to
Aristotelian celestial mechanics, each planet and luminary
travels on a sphere concentric with the Earth. But none of the
Peripatetics, including that Eudoxus whom Plato criticized for
doing carpentry instead of geometry, could "save the phenomena" consistently with the concentric principle. The obvious
explanation of the notable alterations in brightness of the
Moon and the planets is their changing distance from Earth. So
mathematicians dismissed Aristotle's spheres in favor of non-
concentric circles on which to mount celestial objects.

Ptolemy's *Almagest* saves the phenomena by displacing the
Earth from the center of the Sun's circular orbit, thus representing the "first anomaly" (which arises from the ellipticity of the
Earth's orbit) as an effect of perspective (see Figure 3a). Since
the Moon and planets also have a first anomaly, the circles

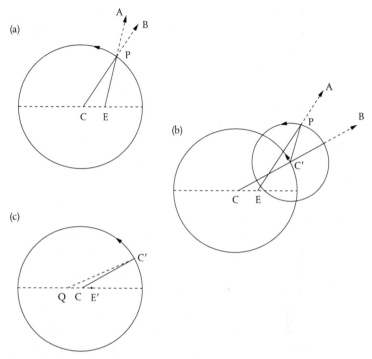

Fig. 3. Ptolemaic bric-a-brac. (a) Saving the first anomaly: P, the planet orbiting at constant velocity on a circle centered at C ("the eccentric"); EA, CB, directions of P as seen from E and C, respectively. (b) Saving the second anomaly: C′, the center of P's epicycle, moves cc (counterclockwise) on eccentric C as P moves cc around it, both motions at constant velocity. (c) The equant: C′ moves cc with constant velocity around Q ("the equant point") where QC = CE′ = CE/2.

representing their orbits (the "eccentrics") also have centers displaced, though in different directions and magnitudes, from the Earth's. To save the "second anomaly," retrogradation, during which a "superior" planet (Mars, Jupiter, or Saturn) appears to reverse directions periodically in its orbit around the Earth, the *Almagest* plants the planet on an "epicycle" turning at constant speed around its center, which rides on the planet's eccentric (see Figure 3b). The combination also explained the change in brightness, provided that the motions were so tuned that retrogradation occurred in mysterious synchronization with the motion of the Sun. A splendid refinement, mimicking the first anomaly almost perfectly, made the center of the epicycle rotate with constant velocity around an "equant" point placed as far on one side of the center of the eccentric as the Earth was on the other (see Figure 3c).

This brilliant, complex bric-a-brac violated good physics by postulating revolutions around unoccupied points without giving any physical reason for their position or motion. They were fictions introduced for description, not explanation. Ptolemy was not content with the division of labor by which *physici* aimed at the truth about substance and mathematicians at description of its accidents. Reversing the usual precedence, he declared the priority of mathematics over physics in the *Almagest* and exemplified it in the astrophysics of his *Hypotheses of the planets*. Here he made the Sun run in a groove between two concentric spheres whose common center is offset from the center of the Earth. Each superior planet rides on a marble (a reified epicycle) running in a similar groove between another pair of eccentric spheres. Accepting the Aristotelian (and Stoic) prohibition of a vacuum, Ptolemy close-packed his system so that the apogee of one celestial body coincides with the perigee of the next, and the greatest distance of Saturn lies in the vault of the stars. Knowing from eclipse observations the average lunar distance in terms of the Earth's radius r and, from his astronomical calculations, the ratios of the radii of the epicyles and eccentrics of each planet, Ptolemy could work up through his close-packed system to find the distance to the stars, which he made out to be just shy of 20,000 terrestrial radii (r).

This number, though vastly less than the true radius of Saturn's orbit, was large enough to suggest the insignificance of human kind, especially when expressed in miles. Eratosthenes of Cyrene, one-time head of the Library of Alexandria, converted r to human measure by determining the angle α between the vertical and the direction of the Sun's rays at Alexandria at noon at the same instant that it stood overhead at Aswan, 5,000 stadia due south. Since α equals the difference in latitude between the two places, Eratosthenes could calculate, from $2\pi r{:}5{,}000 = 360°{:}\alpha$, that $r \approx 4{,}000$ miles (see Figure 8). The radius of the visible world would then be 80 million miles. These numbers were scarcely bettered before the seventeenth century.

Among other enduring pieces of mixed mathematics from the Hellenistic period are Archimedes' demonstration of the law of the balance (in equilibrium, a weight's distance from the fulcrum is inversely proportional to its magnitude), his hydrostatics with its associated concept of specific gravity, an anonymous treatise on mechanics incorrectly attributed to Aristotle, and parts of texts on optics by Euclid and Ptolemy. The optical works, which are mainly geometry, presuppose at least three physical principles of interest: the rays responsible for vision originate *from* the eye, change direction abruptly at the surface between media of different density, and take the shortest path between eye and object. Thus the angles of incidence and reflection are equal.

In a treatise perhaps written by Ptolemy, the author offered experiments to confirm the law of reflection and the breaking of visual rays when they strike an air–water surface. The method gave results that differ insignificantly from Snell's law for angles of incidence up to 40°. The Ptolemaic author observed that refraction must also occur at the boundary between air and ether (or fire or quintessence) and, although he could not calculate the amount, rightly stated that the effect makes celestial objects near the horizon look higher in the sky than they are.

The pseudo-Aristotelian *Mechanica* exceeds the scope of *physica* in being addressed to mechanical problems. But it begins with some physics talk about the marvels of circular motion, in which contraries exist simultaneously. In a spinning wheel, the highest point moves forward and downward as the lowest point moves backward and upward; and the rim, though larger than the hub, moves faster. Pseudo-Aristotle explains that the shorter the spoke, the closer its extremity to the stationary center. By an assumed principle of continuity, the proximity inhibits the motion. On the principle that longer radii are moved more easily through a given angle than shorter ones, the author delivers the law of the lever and deftly applies it to oars, rudders, sails, wheels, pulleys, capstans, wedges, tooth extractors, and nutcrackers. From the nutcracker he proceeds to the head-cracker later known as "Aristotle's wheel." Two circles fixed together concentrically roll horizontally. Although the larger lays down a longer path than the smaller, they stay together. Why? "It is strange . . . and wonderful." The wonder rolled all the way to Galileo, who took many hints and problems from the *Mechanica* believing that they came from Aristotle.

Ancient applications of what today would be deemed engineering physics included the hydraulics of the magnificent system of aqueducts, tunnels, and fountains that supplied Rome's water. "There was never any desseine in the whole world enterprised and effected, more admirable than this," so said Pliny, the expert on marvels. The ancients also knew of the suction pump and waterwheel but made little use of either. Perhaps the best known of their waterworks apart from aqueducts push water around for theatrical effects. The greatest impresario in this line, the military engineer Hero of Alexandria, opened his treatise on ingenious devices with proofs that air is corporeal and that, in union with the other elements and principles, it can produce useful and amazing effects. To produce amazement, place two figures before a hollow altar on a pedestal filled with wine. Conceal hollow tubes inside the figures and the wine-filled pedestal, and run them from the interior of the pedestal to bowls the figures hold. Then kindle a fire on the altar. The air within it expands, pushing the wine in the pedestal into the bowls and thence, as a libation, into the fire. Alternatively, use the pressure to open a door mysteriously (see Figure 4). Such games did not lead to steam engines.

Dumbing down

Owing to disinterest, disturbances, or the rise of Christianity, Latin speakers gradually ceased frequenting Greek schools, and educated Greek slaves, who taught their language to many an aristocratic family, became scarce. Knowledge of Greek, and the copying of Greek texts, dwindled, to arrive functionally at zero

Fig. 4. Sabbath machinery. Air heated by the fire on the altar pushes water and steam from the vessel H into the pot M, which, in descending, pulls ropes that cause the axles they encircle to rotate and open the temple door.

by the time the Ostrogoth Theodoric became King of Italy in 493. Curiously, Theodoric knew more Greek than his learned Latin subjects, having been deposited as a boy in Byzantium as a hostage for the good behavior of the Goths who lived on its borders.

Christian education beyond the Latin schools that taught skills useful to orators and bureaucrats was haphazard and, in the higher reaches of biblical interpretation, largely self-taught. That might help account for the breadth of opinion, and frequency of heresy, among the early self-made theologians. Many of them knew the Latin classics and the writings of the Fathers East and West. They conveyed snippets of the *physica* they knew in their Bible commentaries, particularly on the six days of creation (Hexaemeron), and in general the Bible offered the energetic exegete a wider range of physical subjects than the school texts did the grammarian. However, Augustine's call in his *De doctrina christiana* for a dictionary of ancient science to assist understanding of Scripture met with only partial success.

Lacking such a text in the fifth century, Latin admirers of *physica* resorted to the many compendia copied or composed during the time of troubles. Of first importance was Pliny's *Natural history*. Drawing on him and others, Macrobius Theodosius, perhaps a Greek-speaking bureaucrat in the service of Theodoric, sugarcoated the liberal arts in a *Commentary on the dream of Scipio* that would suit the taste of the Middle Ages. In the dream on which Macrobius hangs his learning, Cicero describes a visit of the deceased Scipio Africanus, the scourge of Carthage, to Scipio's grandson. Africanus reports that the souls of true statesmen dwell in the Milky Way between their incarnations on Earth, and he transports young Scipio there to compare the vastness of the heavens and their harmonious revolutions with the pettiness and discord of the greatest terrestrial empires.

Proceeding downwards from the soul emporium of the Milky Way, Macrobius defines the equator, horizon, meridian, and colures, the ecliptic and eclipses, the zodiac and the solstices; describes the apparent motions of Sun, Moon, and planets up to the first anomaly; and surveys the Earth's five climes, impenetrable torrid zone, and surrounding ocean. Though impressive to its inhabitants, Earth, being but a speck compared with the heavens, is easily sustained in the world's center by the pressure of air exerted equally on all sides. Thus suspended, its inhabitants can appreciate the reason that the Demiurge gave planets and luminaries their particular distances and periods. They make heavenly music keyed to the intervals Pythagoras discovered by plucking lute strings on Earth. If Macrobius knew that these intervals differed altogether from the planetary parameters calculated by Ptolemy, the discrepancy did not bother him. He referred quantitative details to people "disengaged from serious matters."

While descending from the Milky Way to its temporary abode on Earth, the pure soul acquires the properties it will need from the planetary spheres: reason from Saturn's, power to act from Jupiter's, boldness from Mars', sense perception and imagination from the Sun's, passion from Venus', speech and analytical capacity from Mercury's, and, from the Moon's, the final preparation, the ability to perform the physical functions incident to life among the dregs of creation. Macrobius conveys much interesting misinformation in describing this journey: for example, that the planets shine by their own light, the Earth does not reflect sunlight, and Venus and Mercury circulate between the Sun and Mars. Giving Plotinus as his authority, Macrobius makes the planets and luminaries merely the signs, not the movers, of future events. The soul has free will and Africanus urges Scipio to make good use of his, live the moral life, and insure that his soul returns to the Milky Way.

Macrobius' blend of *physica*, Neoplatonism, numerology, metempsychosis, moralizing, and misinformation was not the only vehicle of transmission of ancient texts produced by high-level civil servants (if such he was) in and around the court of Theodoric. For some time Theodoric's chief advisor was the Catholic Boethius, scion of a high patrician family, whose old-fashioned education gave him fluent Greek, and whose concern that philosophy might be lost prompted him to undertake a Latin translation of all of Aristotle. He began with the *Organon* (Aristotle's logical treatises), which was immediately useful in head-to-head religious controversy and so could serve as a Trojan horse for the rest of Peripatetic philosophy. Boethius did not complete much of his project, however, before Theodoric, believing him to be a traitor, had him killed. While awaiting execution, Boethius wrote his famous *Consolations of philosophy*, in which a Christian Neoplatonic world view gave him the resolve to face his misfortune stoically. Together with Aristotle's logic and the application of high philosophy to consolation, Boethius bequeathed to medieval letters a short tract on music, which covered the terrestrial and celestial types, and the harmony of the soul. He was succeeded as Theodoric's minister by a lawyer, Cassiodorus, equally committed to preserving ancient learning but in a form more useful to Christians than translations of Aristotle. He set up a monastery for whose inmates he rendered the seven liberal arts in a manner suitable to Christians struggling to preserve some remnants of Roman culture in an increasingly barbarous world.

The most popular rendition of the seven liberal arts bequeathed to the Latin Middle Ages was a work of the fifth century, the *Wedding of Mercury and philology*, of whose author C.S. Lewis remarked that "this universe, which has produced the bee-orchid and the giraffe, has produced nothing stranger than Martianus Capella." Martianus set forth what he knew about the liberal arts in lectures by philology's bridesmaids. They deliver their wisdom in excerpts from ancient primers or enunciations of propositions without proofs. Much of what they say is unintelligible. Thus Astronomy, having stepped from her ball of heavenly light, declared that during the 40,000 years she spent in Egypt studying her subject she learned that celestial natures, "circling by their own surging, are diffused the entire way around in globular belts and circles." After this venture into *physica* she described the apparent motions of the planets and luminaries, defined the equator and zodiac, tried to explain how to measure the Earth, climbed back into her ball, and, like the antiquity she represents, flew away.

Fig. 5. Lecture in a House of Wisdom. The site is the library in Basra, the time the thirteenth century.

2

SELECTION IN ISLAM

———◆·◆·◆———

Among the Christians who equipped themselves with
Aristotle's *Organon* to dispute about the nature of the
Trinity and other enigmas were followers of the patriarch of
Constantinople, Nestorius, who was excommunicated in 431 CE.
When expelled from the Roman Empire they settled in Persia, at
Jundishapur, a town built 200 years earlier to house captive
Roman soldiers. There the Nestorians built an important school
and hospital in which they enlarged their study of Aristotle and
Galen. Vigorous missionaries, the Nestorians became experts in
translating difficult texts between unrelated languages. The early
Arabian conquests to the east increased the scope of Nestorian
missionary activity. The caliphs usually considered them as allies
because they had enemies and beliefs in common, and the
Qur'ān recognized them as fellow "People of the Book," provided they paid their taxes.

The first era of Arabian expansion ended in the mid-eighth
century. The dynasty that had presided over most of it, the

Umayyads, headquartered in Damascus, succumbed to a clan with Persian backing, the Abbasids, who built a new city, Baghdad, as its power base. Nestorian learning found a growing audience in the eclectic urban elite drawn to the new capital from around the conquered lands. In 765 the director of the hospital at Jundishapur became the Abbasids' court physician. Syriac-speaking Christian doctors soon swarmed in the new capital. They supplied the translators who made Aristotle "The Philosopher" in the Arabic language.

The project to render Greek science into Arabic lasted three centuries. The translators had to invent many new technical terms, even a word for philosophy, for which they adopted the transliteration *falsafa*. In the early ninth century, the Abbasid Caliph al-Mamūn (813–33) established an academy or research center and library at Baghdad, the House of Wisdom, which gathered works by Greek medical writers and mathematicians as well as Aristotle. By 1050 every Aristotelian treatise except the *Politics* existed in Arabic, together with many books like the *Mechanica* incorrectly ascribed to him (see Figure 5).

The wealthy of Baghdad supported this activity, which, during the heyday of the House of Wisdom, centered on a wide competitive search for ancient manuscripts. This unusual treasure hunt had to do with the rapid rise of the Islamic state. Lacking administrative experience, the Arabian conquerors employed the *dīwān*, or bureaucratic offices and procedures of the Persian Empire, to tax, survey, and keep records. The new empire kept its books in Persian until late Umayyad times when it shifted to Arabic. That required translation of the *dīwān* manuals, which probably included introductions to arithmetic, geometry, and branches of applied mathematics.

The shift to Arabic opened the bureaucracy to place hunters and, insofar as careers were open to talent, intense competition. Mastery of applied mathematics may have helped people with ambitions of upper-level appointments; hence the hunt for mathematical treatises in depositories in Byzantium and elsewhere. Once domesticated by this process, mixed mathematics became a pursuit in its own right, though still supported for its applications to administration, engineering, and religious observance. Thus, unlike the Greeks, most Muslim *physici* and mathematicians were courtiers, officials, or functionaries.

The arts of reasoning and persuasion, useful for making headway in Baghdad's sophisticated philosophical salons and a name in religious controversy, again served as a stalking horse for the rest of Aristotle's system. The early Abbasids favored a liberal form of *kalām* (scholastic exegesis of religious writings), *Mu'tazila*, which allowed wide scope to reason in the interpretation of obscure or implausible passages in Scripture and in the attainment of moral as well as philosophical truth. The Aristotelian corpus showed what reason unaided by revelation could achieve. It also bore directly on such practical matters as astrology and meteorology, and hylomorphism (the doctrine of form and matter) provided the vocabulary for discussion, if not the ingredients for solution, of most questions about everyday experience.

These practical considerations may be taken as the efficient cause of the translations and the commentaries and compendia descended from them. The principles of knowledge, and the ways to wisdom, might be adduced as formal and final causes. There remains the material cause. It was paper. The Arabs became acquainted with this Chinese invention during their conquests in Central Asia. A paper mill operated in Baghdad before 800 to help meet the needs of record keepers. More paper became available just when al-Mamūn's translators needed it. A large market in books developed, supplied by an assembly line, or rather circle, of scribes who simultaneously took dictation from an author or scholar. The scribes read back what they had taken down and the dictator, if satisfied, "authorized" the copies. The system produced more, and more faithful, versions of the originals than monastic scribes could make one at a time on vellum or parchment. Colossal libraries came into being, the largest, at 400,000 or 500,000 volumes, being over a hundred times the size of their nearest competitors in the West.

Central administration of the new empire proved impossible. As soon as the great expansion peaked in the eighth century its parts began to separate. In the tenth century, caliphates split off in Egypt under a dynasty claiming direct descent from the Prophet's daughter Fatima, and in Andalusia under a branch of the Ummayads. At the same time, the Abbasids lost all power in the East to Turkish and Persian shahs and emirs. These arrangements proved fleeting: in 1031 the Umayyads lost Andalusia to fundamentalist Muslims from North Africa; in 1171 the Fatimids gave way to the Kurd Saladin; in 1258 the Mongols destroyed Baghdad. Nonetheless, a common language for literature and administration, and conversions to Islam, kept the disparate political elements together as a cultural, religious, and commercial "empire."

Because of the wide distribution in place and time of niches for its cultivation, *falsafa* could develop only in fits and starts in Islamic lands. Hence the enormous energy devoted to producing digests, encyclopedias, and commentaries on standard material by the most gifted of the Islamic philosophers, and a reason that they did not develop a *physica* much different from what they inherited. Another reason is that, in contrast to mathematics, *physica* had consequences for faith. Aristotle's Indifferent God, whether pure or, as was largely the case, Neoplatonized, could be Islamized, as he was Christianized, only at the expense of the logical consistency of the system. To insist on pure Aristotle, with his eternal world and perishable soul, invited theological opposition.

Falsafa

The House of Wisdom established in Baghdad by al-Mamūn incorporated the library of his father, Harun al-Rashid, the Caliph of the *Thousand and one nights,* and an observatory for other nocturnal activities. It also sponsored expeditions to measure the size of the Earth. Like the House's translation project, its astronomical and geographical observations arose from Greek prompts. Al-Mamūn wanted to check Ptolemy and Eratosthenes for the surer estimate of the size of his empire and the relative positions of the principal places in it. Geography flourished wherever encouraged and, with history, made the largest body of literature in the Islamic languages during the Middle Ages, apart from religion. The great figure of the House of Wisdom was not a geographer, however, but an exemplary polymath, the founder of Arabic Aristotelianism, al-Kindī. He is a good example of the generalization made earlier about the social status of Muslim savants. His father, from Yemen, was governor of Kufa in Iraq, whence al-Kindī made his way to Baghdad. In his version of Aristotle, the visible world consists of the four restless elements, the lazy circulating quintessence, form, matter, substance, and accident, and runs on the four causes. The Unmoved Mover transmits motion, and brings about substantial change, through the rotation of the heavens. This stationary Being is Neoplatonic in holding in its mind the Ideas and Forms of created things and Qur'ānic in knowing the particulars of the physical world it created and the characters and deeds of everyone alive and dead. With this adjustment, al-Kindī's world picture became the standard model of Aristotelian *physica* until the twelfth century, when defenders of *falsafa* in Spain advocated a stricter reading of the Philosopher.

At first the amalgam fared well, especially in the hands of al-Fārābī, the son of a Muslim general. Born in remote Turkestan and schooled by a Nestorian Christian, he arrived in Baghdad to earn the sobriquet of "Second Teacher," Aristotle being the first, for his presentation of Peripatetic philosophy improved by the insights of Islam. His ideal syllabus, *The attainment of happiness,* advises the truth seeker to begin with things easiest to understand—numbers and geometrical figures—and proceed gradually toward the material world. The journey leads through optics, astronomy, music, and mechanical principles taken as Archimedean abstractions to *physica*, the science of the things that make up the world. Proceeding again from the least material to the most, the pursuit of happiness sets out from the heavenly bodies and descends through the four elements to stones and the Earth's interior.

Having understood the four causes of all the species and actions of the physical world, the inquisitive mind inquires about beings more perfect than nature and natural things, and sublimes to metaphysics. Beginning with the soul and intellect of the rational animal, it discovers the way of human perfection and climbs back up the chain of principles, peeling away the material aspects, ascending through ever more perfect rational creatures to the First Principle, the Being too perfect for description. The fulfilled mind now understands the nature and place of everything in the universe, including the Islamic state: for the caliph relates to his hierarchy of subordinates down to the lowest citizen as the First Principle relates to created beings from the highest intellect to the densest stone.

This clear teaching and the support of the Abbasids did not suffice to establish a new Lyceum in Baghdad. Neither al-Kindī nor al-Fārābī led continuous academic lives in the capital. Al-Mamūn's effort to impose *Mu'tazila* failed and his successors oppressed its representatives. The House of Wisdom almost perished and al-Kindī had to leave. Al-Fārābī retained the support of his caliph, with the unhelpful consequence that they were driven out of Baghdad together. After the Second Teacher died eight years later in Damascus, *falsafa* moved far into Persia, where in 980 it produced one of its finest flowers, Ibn Sīnā (Avicenna), in Bukhara, in what is now Uzbekistan. Extravagantly precocious, he drifted around Persia in search of knowledge, often serving as a physician to the reigning power. His capacious memory held the entire Qur'ān, and much of Greek medicine, mathematics, and *physica*, which he poured into an antidote to ignorance and insomnia entitled *The cure*.

The cure proposes the same project as al-Fārābī's *Attainment of happiness*, but arranges the steps in the order of the Aristotelian canon. Thus after the *Organon* come the basic principles of *physica*, their applications to the celestial and sublunar realms, the mysteries of motion and the human mind, and then mathematics. The last topic is First Philosophy, culminating in knowledge of the One who operates in the Neoplatonic manner of emanation and delegation. Seeking the unification of all knowledge and practice under one set of principles sanctioned by the Qur'ān's insistence on the oneness of God, Avicenna took on human activities (prayer, prophecy, politics, law) as well as the natural and supernatural worlds, and the techniques, the logic and mathematics, necessary to study them.

Avicenna's widely read works ended, or suspended, the productive cultivation of *falsafa* in the eastern reaches of Islam. That was owing primarily to al-Ghazālī (Algazel), a conscientious theologian well versed in *kalām* and *falsafa*. After isolating himself for some years to ponder the relations between reason and revelation, he announced that *kalām* could resolve nothing important for faith, and that *falsafa* was inimical to it. He emphasized the particulars in which the Unmoved Mover differed from the God of Islam, and pointed out revealed details missing from *falsafa*, such as the Last Judgment, the resurrection of souls, and the dissolution of the world. He allowed the practice of *kalām* where helpful in persuading wavering believers of a rationalistic tendency; but, in general, he thought theology and the Sufism to which he inclined would do better without it. His position had the strength of logic and of a new Seljuk ruler in Baghdad, who favored a stronger orthodoxy than the more relaxed religion of the Abbasid caliphs.

The last stage in Aristotle's journey through Islam jumped two continents, from Persia to Spain, and over a century, from Avicenna to Ibn Bājja (Avempace), the long-serving vizier of the governor of Granada, and Ibn Rushd (Averroes), one-time *cadi*, or religious judge, in Seville and Cordoba. Both insisted on a purer Aristotle than al-Kindī's, although with reservations. The better known of the two, Averroes, made his career as a physician and protégé of the Almohad Caliph Abd al-Mu'min before becoming *cadi*. His literalist renderings of Aristotle and his opposition to al-Ghazālī's teachings made him the target of traditionalists worried by the successes of the Christian reconquest of Spain. Eventually a high court in Cordoba condemned his approach.

The problems raised by the schism between astronomy and physics appear to have worried Averroes for most of his philosophical life. As a preliminary to their resolution, he tried to separate Aristotelian cosmology from the Neoplatonic and Islamic accretions it had acquired since the time of al-Kindī. That this mélange had become the default *falsafa* appears from the fable of "Hai Eb'n Yockdan," as its hero was named in its English translation. Abandoned on a desert island as an infant, Yockdan had plenty of time to think about nature and his place in it. Step by logical step, he invented a Neoplatonic universe on which, after instruction by a passing holy man, he successfully grafted Islam. Averroes declared war on Yockdans. In perceptive analyses of the Aristotelian corpus that would earn him the title of "The Commentator" in the Latin West, he removed the chain of creating and created beings between the One and the lunar Intelligence, restored the eternal cosmos and the Indifferent Mover, severely criticized both Avicenna and al-Ghazālī, and allowed that Aristotle had possessed as much of the truth as a man can obtain without revelation. One of these truths was that Ptolemaic astronomy perpetrated a sham: it might be good mathematics, but it had nothing to do with the real world. The Commentator thus left the Aristotelian corpus as the Arabs had found it 400 years earlier, so that it might almost be said of the gigantic effort of *falsafa* what Omar Khayyam said of himself after hearing the saints and doctors dispute: "[I] evermore came out by the same door as in I went."

The blunting of *kalām* and discouragement of *falsafa* coincided with an important change in the Islamic educational system. The Abbasid House of Wisdom, with its library and observatory, and its even grander reproduction in Cairo by the Caliph al-Hakīm around 1000, were the high end of a type reproduced in several places in Iraq and Fatamid Egypt. Most in fact were little more than libraries, though some offered ink and the ubiquitous paper to their readers. Originally given over to the "foreign sciences," the libraries either died out or converted to cultivating the "Islamic sciences" of law, religious studies, and Arabic philology. They did not constitute an educational system.

In early Islamic times, regular elementary education took place in the mosques. But from the eleventh century, schools founded by traditionalists under a provision (*waqf*) that allowed the endower of a public institution to retain family control over it after his death took over. Mainly Sunni institutions, these *madrasas* excluded the foreign sciences that had flourished in the largely Shia, Persian-inspired, libraries. The great conquerors in the East, Hūlagū Khan, Timur, the Seljuks, were great founders of *madrasas*, which helped to inculcate a standard faith and reduce religious controversy. Teaching relied on staggering feats of memorization, first of the Qur'ān, then of the sayings of the Prophet (*hadith*) and the foundations of the law (*fikh*). It is said of one overachiever that he died having dictated only 30,000 folio pages of the texts he knew by heart.

Mixed mathematics

Astronomy

Astronomers arrived at Averroes' view of Ptolemy's constructions by a steeper path. They began, as did the philosophers, by mastering the translations prepared by denizens of the Baghdad House of Wisdom. That included checking the parameters transmitted in the *Almagest*. After eight centuries, shortfalls between Ptolemy's predictions and observation had become conspicuous. For 150 years or so, until around the year 1000, astronomers working in the Islamic East labored to improve the parameters and the methods by which Ptolemy had deduced them. Thus humankind came to have good values for the inclination of the ecliptic, the position of the Sun's apogee, the precession of the equinoxes, and other desiderata.

Two social factors underpinned this steady progress. The eccentric astronomer Ibn Yūnus, who served the Fatamids and died faithfully in 1009 on the day he had calculated would be his last, specified one of these factors: "observation of the stars agrees with religious law, for it allows us to know the time of prayers and of the sunrise and sunset that mark the beginning and end of fasting." Hence Islamic astronomers investigated assiduously topics of only passing interest to Ptolemy, like the duration of dawn and dusk, and conditions for glimpsing the first appearance of the new Moon. Also, astronomy taught the direction to Mecca, the *qibla*, from any place assigned, when to plant, and how to get from one place to another for commerce, pillage, or pilgrimage. Thus, in contrast to *physica*, the study of the stars, gently recommended in the Qur'ān (6.97, 10.5, 16.12, 16, 71.15–16), had the support of religious leaders. It also had the support of secular rulers, not only because of its help in imposing religious conformity and in guiding caravans, but also because of its foundational role in astrology. Ibn Yūnus was an expert on the subject and his primary patron, the Fatamid Caliph al-Hakīm, the builder of Cairo's House of Wisdom, was devoted to it. Ibn Yūnus's huge *zij*, or handbook of astronomical information, contains tens of thousands of bits of astronomical-astrological data.

The second social factor in the progress of Arabic astronomy was continuity. Whereas *physica* convulsed periodically, astronomical observation went on almost continuously, in two different senses. For one, observatories, even if they did not outlast their founders, supported frequent determinations of the positions of the luminaries and planets, whereas Ptolemy had calculated his parameters from very few strategically timed observations.

Astronomers designed observational programs to last for an entire circuit of Jupiter (12 years) or, even more optimistically, of Saturn (30 years). The Islamic lands were fruitful in astronomers, *zijēs*, and monumental instruments. For example, at Rayy, 12 kilometers south of Teheran, al-Khujandī, who died in 1000, built a sunken sextant with a radius of 20 meters made of wood sheathed in copper and graduated to ten minutes of arc. With this giant, paid for by the local Persian strongman, he made exact determinations of the obliquity of the ecliptic and the latitude of the instrument.

By 1050 Ptolemaic astronomy had reached its apogee. Astronomers had corrected its parameters via observations with substantial instruments, invented new ways to deduce the parameters from observations, simplified its mathematics with plane and spherical trigonometry, and replaced the cumbersome ancient notation with "Arabic" (that is, Hindu) numerals. What more was there to do? The approaches of three Persian mathematicians of the mid-eleventh century—al-Bīrūnī, al-Khayyāmī (Omar Khayyam), and Ibn al-Haytham (Alhazen)—will indicate the options available.

Al-Bīrūnī, whose erudition was extraordinary even for a Muslim scholar, suffered in good measure the peripeteias of Islamic savants dependent on courts. During his lifetime the north-east fringe of the territory claimed by the powerless Abbasid caliphs changed hands several times. Born and educated in Khwārazm, he fled civil war to seek patronage in the fragmenting kingdoms around him. He met al-Khujandī, observed at the great Rayy sextant, and found temporary employment with fleeting strongmen, one of whom built him an observatory.

After the murder of this Maecenas, al-Bīrūnī continued his astronomical and geographical observations in and around Ghazna (Afghanistan). The powerful Ghaznavid Sultan Mahmud knew so little about astronomy that he rejected as heresy the assertion that in the far north the Sun sometimes does not shine for days. Al-Bīrūnī enlightened him and Mahmud reciprocated by giving al-Bīrūnī the opportunity to become the sage of the age. Following the Ghaznavid armies to the east, he mastered Sanskrit, determined geographical positions, and brought back much miscellaneous Indian lore.

Most of al-Bīrūnī's writing concerns astronomy, geography, and geodesy. Although Ptolemaic in conception, his astronomy analyzes other views, for example, the possibility that the apparent motion of the stars arises from a rotation of the Earth. He gave an Indian as well as a Greek source for this idea and reported Ptolemy's objections to it (bodies dropped from a tower would land west of its foot) and the response (all bodies on a rotating Earth would participate in its motion even when falling). As many others would do, al-Bīrūnī accepted Ptolemy's objection, although an incident he recorded suggests that forces beyond his predilections and the demands of his patrons may have recommended his acceptance of the traditional view. He had shown an instrument for finding the time of prayers to an orthodox legalist. The man decried its use because it bore Byzantine names of the months and so opened an entry for the infidel.

Khayyam also suffered the vicissitudes of the Islamic savant for whom good fortune, "like snow upon the desert's dusty face," never lasted long. Born in Khurāsān shortly after the Seljuks conquered the province, he spent his most productive years serving Sultan Malik-Shāh in Isfahan. The sultan and his vizier supported an observatory, where, under Khayyam's direction, a group of astronomers issued their own *zij*. He or they also developed a new calendar with a tropical year closer by three parts in ten million to the truth than the Gregorian rule. The idyll ended with the death of Malik-Shāh and the assassination of his vizier. The new regime halted the calendar reform, beggared the observatory, and welcomed the orthodox, who opposed Khayyam's Avicennan world view and poetical free thinking. The line of astronomy he represented, which sought progress in the next place of decimals and required ongoing dependable financing, thus came to an end in Isfahan.

The third example, Alhazen, took the bold position, later pushed by Averroes, that Ptolemaic astronomy had to be reworked to conform to physical principles. He seems to have been a man of unusual confidence, since he proposed to the Fatamid Caliph al-Hakīm—the patron of Ibn Yūnus—a plan to regulate the flow of the Nile. When the plan failed, Alhazen thought it advisable to feign the insanity that his wild project suggested. He recovered his wits when al-Hakīm died and directed them to optics as well as to the shortcomings of Ptolemy. Insisting that the rules of astrophysics required that every rotation ascribed to celestial objects had to be performed by a material shell or sphere turning with constant velocity around its own center, and that no void could exist in the heavens, Alhazen reified all the deferents and epicycles of mathematical astronomy as so many nested globes and marbles, as portrayed in Ptolemy's *Hypotheses*. This produced only a qualitative version of a physical astronomy, as it lacked a replacement for equant motion.

The three directions of astronomy represented by al-Bīrūnī, Khayyam, and Alhazen came together briefly and dramatically in the work of Nasīr al-Dīn al-Tūsī. From a family of Shia jurists, al-Tūsī received a full Islamic education from several masters, including a follower of Avicenna. Political turmoil in his home-land (like Khayyam he came from Khurāsān) discouraged the life of the mind, so al-Tūsī accepted an invitation to work in the fastness of a tough Shia sect that specialized in murder. In the company of these Assassins and their grand master, "the Old Man of the Mountains," he wrote many important tracts, some on ethics, which they certainly needed, others on logic, philoso-phy, and mathematics. After a quarter century in his unusual academic setting, al-Tūsī upgraded to the service of a grandson of Genghis Khan, Hūlāgū, who destroyed the Assassins in 1256, conquered Baghdad and terminated the Abbasids in 1258, and established his power from the borders of the Byzantine Empire to the fringes of China.

Though a little rough, Hūlāgū had an interest in astronomy and astrology, and encouraged al-Tūsī to gather up manuscripts before his less learned followers ate them. (Perhaps Hūlāgū owed his civility to the Nestorian Christians, from whom his father had chosen his mother and he his favorite wife.) Hūlāgū erected an observatory for al-Tūsī in the new Ilkhan capital of Marāgha in Azerbaijan. This institution, which Hūlāgū not only paid for (which was to be expected) but also, exceptionally, endowed, attracted several excellent astronomers, boasted a library and a librarian, and housed several large instruments including a mural quadrant and an armillary sphere. One of its first prod-ucts, the result of 12 years of observation and calculation was, of course, a *zij*. Al-Tūsī and his collaborators disliked the Ptolemaic models on which they had to base their calculations, and, taking their manifesto from the Aristotelian *physica* eclipsed since al-Ghazālī's victory over Avicenna, developed new planetary models. This departure, imitated in the fourteenth century by Ibn al-Shātir of Damascus, a liturgical technician (he kept track of religious time), turned out to be a useful step, when, using the same models, Copernicus took a greater stride.

Physics

With some imagination, a savant can find verses in the Qur'ān (15.16, 16.20, 24.35) that encourage the study of astrology and optical phenomena. The greatest investigator of light during the Islamic Middle Ages was Alhazen. As in his astronomy, so in his optics but to better effect, he attempted to combine Aristotle's *physica* with geometrical models. Thus, in contrast to most mathematicians, he ascribed vision to rays from a luminous body entering the eye (intromission) rather that to rays from the eye falling on the body (extramission). He distinguished between primary light (from self-luminous bodies like the Sun, planets, and fire), secondary light (from all points on an opaque or transparent body illuminated by primary light), reflected light, and refracted light. This allowed him to state that the Moon shines by secondary, not reflected, light, which rebounds from a polished body only at the angle of incidence—a mathematical version of Plutarch's description of the lunar surface.

According to Alhazen, rays proceed rectilinearly in all directions from every point on a body shining by primary or secondary light. How then can all the forms and colors they bring to the eye be received without confusion? He answered that rays striking the surface of the crystalline humor perpendicularly dominate the image. As for reflected and refracted light, he was content to define the directions of their actions by experiment. He confirmed in detail the equality of the angles of incidence (i) and reflection, and gave qualitative rules relating i to the deviation $d = i - r$ suffered in refraction, r being the angle of refraction.

Many mathematicians who wrote on astronomy also wrote on mechanics, notably, in the Baghdad school, al-Khwārizmī, whose name inspired an English common noun ("algorithm"), and Thābit ibn Qurra, a money changer from Harrān who became a master of all the mathematical sciences; and, among the Eastern astronomers, al-Bīrūnī, Khayyam, and Avicenna. They built on Archimedes' theories of the lever and floating bodies, Aristotle's concepts of weight and motion, and pseudo-Aristotle's analysis of simple machines (*Mechanica*). The essential interest of their work lies in its comparative realism: they endeavored, more successfully than in their astronomy, to unite the physical and the mathematical. Thus they did not neglect the weight of the beam when analyzing the equilibrium of a balance or the operation of a lever; and they invented or perfected a "balance of wisdom" with which to measure the specific gravity of anything submersible in a liquid without provoking a chemical reaction. In the case of two-component alloys, algebraic manipulation of the measurements evinced the percentage of the ingredients and gave a more convenient way of detecting counterfeits than Archimedes' method of dropping them into his bathtub.

The most detailed account of Islamic statics and hydrodynamics comes in al-Khāzinī's *Book of the balance of wisdom*, which includes an extensive table of the specific gravities of metals and minerals made by al-Bīrūnī. Al-Khāzinī, who wrote between 1115 and 1130, was the well-educated Byzantine slave of the treasurer of the Seljuk court at Merv in Khurāsān. He debuted in the normal fashion of court astronomers, with a *zij* dedicated to the ruler. His main accomplishment was the construction of the most precise of balances, a work inspired in equal measure by duty to his master the treasurer, who worried about counterfeit gems and alloys, and pious curiosity about the Day of Judgment, when, as the Qur'ān frequently advises, our deeds will be weighed to the last scruple. The detailed operation of al-Khāzinī's balance, which employed five weighing pans and reached an accuracy he estimated as one part in 60,000, is less important to know now than his analysis of the stability of balances suspended from axes above, through, or under the beam's center of gravity.

Al-Khāzinī's analysis makes the weight of a body depend on its position in two distinct ways. It is the greater the further it hangs from the axis of rotation, and the less the higher the specific gravity or density of the medium it occupies. Invoking a strict though inappropriate analogy to the "weight" (that is, torque) of a body on a balance arm, which runs from zero on the axis to a maximum at the arm's end, al-Khāzinī made a body's weight a maximum at the concave surface of the lunar sphere and zero at the Earth's center. The concept of "weight according to position" and attendant puzzles such as whether a body has weight in its own milieu, as air in air, still figured prominently in Aristotelian *physica* when Galileo learned it as a student at the University of Pisa.

That was around 1589, just when astrological doctrine elaborated under the early Abbasids by the Jew Māshā' Allāh, who had helped to calculate a proper day for the founding of Baghdad, predicted an important alteration in human affairs. His astral history turned on the recondite ideas of grand conjunction and trigon change. A trigon is a group of zodiacal signs 120 degrees apart; a grand conjunction brings together Jupiter and Saturn at 20-year intervals; after 200 years or so, during which these reunions take place within a given trigon, they move over to another, and in so doing cause great disruptions (see Figure 6). Do the transitions physically produce disasters or only announce them? Although the Qur'ān suggests (15.81) and reason confirms that the heavens merely advertise, some astrologers, who identified each trigon with one of the four terrestrial elements, predicted disasters related to the qualities of the element characterizing the receiving trigon. Thus when Jupiter and Saturn first come together in the watery trigon they very probably will cause a flood. The greatest effect of trigon shifts occurs after 800 years when the conjunctions complete a full cycle. Hence Galileo arrived eight centuries after Māshā' Allāh.

Does astrology work through physical agency and therefore belong to *physica*, or only give notice of future happenings through signs predictable by astronomy, and thus fall to mathematics? In this guise the old problem of the relations between

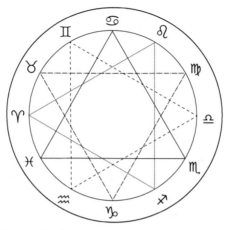

Fig. 6. Astral history. Grand conjunctions occur every 20 years, trigon changes every 200 years, and cycle changes every 800 years, more or less. The solid triangle connects the watery signs, the dashed the airy, the dotted the earthy, and the dot-dashed the fiery.

mathematics and physics acquired the additional complication of whether astrology's predictions, if correctly interpreted, could be avoided. The Qur'ān, which declares frequently that only God knows the future, hints at a possibility of avoidance (16.12, 27.65, 67.5–9). Avicenna and Averroes appealed to these declarations to convict astrology of futility. But the early adepts, like Māshā' Allāh, al-Kindī, and Thābit ibn Qurra, who worked the literal star lore of Persia into Islamic astrology, accepted the fatalistic principles and predictions of their art. The strictest advocate was the leading Muslim astrologer Abū Ma'shar, another Persian from Khurāsān, who arrived in Baghdad close to the beginning of the reign of al-Mamūn. He taught that the only way for human souls dwelling beneath the Moon to return to the One in the region of light above the stars is to enlist the help of the deities of the planets and constellations. To win it the soul must learn the characters of the stars by studying how astrological influences act physically on terrestrial substances.

Further illumination on the question of celestial influence can be found in medical practice. Astrologers associated a zodiacal sign with body parts, thus Aries with the head, Taurus with the neck, the Gemini with the shoulders, down to Pisces and the feet. As anyone can work out by observation of the tides and the menstrual cycle, the Moon has some special influence over fluids. Hence it might cause unwanted interference during medically sanctioned bleeding. To minimize the risk, physicians avoided bleeding from any part of the body associated with the zodiacal sign through which the Moon was passing. This beneficial abstention is more plausibly construed as a response to a risk from a real physical force than to a notice delivered by abstract signs. Hence *melathesia*, the name of this branch of medical practice, was further evidence of the need to integrate terrestrial physics with the mixed mathematics of the heavens.

Departures

In glaring contrast to the philosophers of Greece, members of the Baghdad House of Wisdom from al-Kindī on, and men of many sciences like al-Bīrūnī, Alhazen, and Avicenna, wrote on technological and engineering subjects. The Islamic states undertook large-scale works—for example, the construction of Baghdad, with its mosques, gardens, and palaces—and large-scale wars, with their demands for weapons and fortifications. The decorations of the buildings and their furnishings incorporated geometrical designs of great beauty and complexity, and the strength of the weapons relied on unique and intricate metallurgical processes. The famous swords made of "Damascene steel" combined beauty with utility. The high-end metal industry nourished by military requirements also supplied parts for advanced engineering projects and scientific instruments. Of these latter, the best known is the planispheric astrolabe, which may stand as the emblem or logo of Islamic science.

The finest astrolabes, which consist of several delicately inscribed brass plates contained in a stubby cylinder, were collectors' items in their time as well as in ours. They could be adjusted to show in a plane the places of the Sun and stars as they are observed in the stellar sphere. The operator needed only to turn the open-work "rete" until one of its star points, or the Sun's position on the ecliptic, came above the appropriate circle of altitude or azimuth on the lower plate or "tympan" (see Figure 7). A good astrolabe carries several tympans engraved on both sides for use at different latitudes. The concept of the instrument is Greek, its realization in brass Islamic. The oldest extant example, from Isfahan, dates from the tenth century.

Once set, the astrolabe gives the time of day, the time and direction of sunrise and sunset, and, with additional information, the *qibla*. It can also serve as a compass for navigation. For these reasons it elegantly met the needs of the religion founded by the merchant Muhammad: it gave the times and orientation of daily prayers, guided pilgrims' steps, and helped caravans cross the desert. And there is more. The sighting bar rotatable

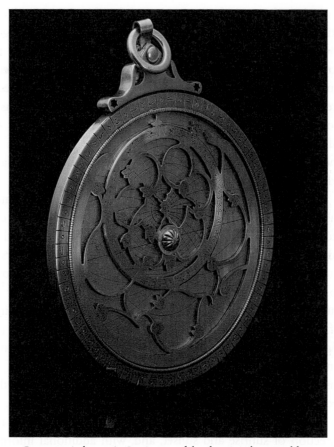

Fig. 7. Computer in brass. An Iranian astrolabe showing the rotatable rete with its star points and, underneath it, a tympan with the fixed reference circles of the observer.

against the circular scales on the back of the instrument can measure the radius of the Earth. Al-Bīrūnī took his astrolabe to the top of a high mountain in India adjacent to a flat plain and measured the angle θ between the vertical and the horizon. Any neophyte in Muslim trigonometry could then work out r from $\sin \theta = r/(r + h)$, h being the mountain height, which al-Bīrūnī measured by a standard surveying technique (see Figure 8). His

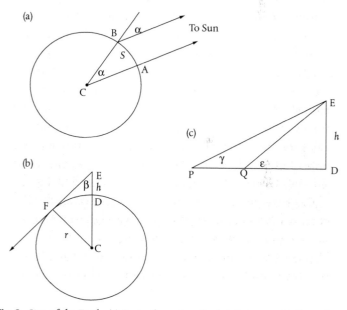

Fig. 8. Size of the Earth. (a) Eratosthenes' method: A is Aswan, B Alexandria, α the measurable zenith distance of the Sun at B, s the separation of A and B as determined by walking, sailing, or guessing. (b) Al-Bīrūnī's method: E is the mountaintop, EF the horizon through F, h the mountain's height; the rest is trigonometry. (c) h determined from measurements of the angles γ, ϵ from the stations P and Q and the paced-out distance between them.

result confirmed the value obtained by al-Mamūn's mathematicians, although "their instruments were more precise, and their labor to obtain it of an extremely exacting and fastidious nature."

We return briefly to the House of Wisdom, where the Banū Mūsā—three sons of a bandit named Mūsā ibn Shākir, who went upmarket to die an astrologer at al-Mamūn's court—diverted themselves from mathematics to improve the devices of Hero of Alexandria. They added some complicated siphons that, among other feats, appeared to discharge wine and water separately after imbibing them mixed. The greatest mechanician in this tradition worked for a small Turkish dynasty that ruled land around the upper Tigris in the twelfth and thirteenth centuries. He was al-Jazarī, whose *Compendium of the theory and practice of the practical arts*, "the most important document on machines from ancient times to the renaissance from any cultural area," describes feedback control and other novel techniques. A fine example of his automatons, the elephant clock, carries an open domed tower with a phoenix on top (see Figure 9). A figure sits on a ledge beneath the dome flanked by falcon heads. Half way down the tower under each falcon a dragon rides on a horizontal axis. A scribe sits at the tower's base just behind the mahout. The elephant's body contains a large vessel full of water in which a perforated wooden bucket, loosely attached by ropes to some gearing, floats. Action begins when, in precisely 30 minutes, the leaky bucket sinks, pulling the ropes and engaging the gears, which make the phoenix whirl, causing a ball to fall and the figure on the ledge to decide from which of the falcon heads the ball will drop into a dragon's mouth. The lucky dragon swallows the ball, swings on its axis, rewinds the ropes, and brings the

Fig. 9. The magnificent elephant clock. An elaborate automaton built by al-Jazarī that celebrates every half hour, day and night.

bucket back to the surface. The clock continues to play every half hour until it has exhausted its balls.

Al-Bīrūnī mentions that gears were applied to astrolabes. The earliest surviving example, which is also the earliest extant intact gear train, dates from 1221. Five finely cut gears mounted inside the backplate drive images of the Moon to indicate its place and age. Dots representing the Sun and Moon show their ecliptic longitudes. Nibs on the rete provide a grip for turning the gears. The inclusion of information about the Moon's zodiacal position made the astrolabe a medical instrument, since it allowed the doctor to discover with a twist whether or not he should cut his patient. If the twist were provided by a falling weight controlled by an escapement, the geared astrolabe would be a mechanical clock. That is exactly what happened in the West in the fourteenth century. Some of these clocks, and many subsequent imitations, still exist, inadvertently advertising, via their astrolabic faces and solar and lunar pointers, the logo and achievements of medieval Muslim science.

Fig. 10. Stoned students. A scene from the arts faculty at the University of Bologna in the thirteenth century, as recorded on a tombstone.

3

DOMESTICATION
IN EUROPE

———•◦•———

The domestication of Greek and Arabic *physica* and mixed mathematics in the Latin West took some 400 years from the first translations in the twelfth century until well into the sixteenth, when Archimedes and Ptolemy were printed and Aristotle's ancient rivals revitalized. With the generation of Galileo, Kepler, and Francis Bacon, *physica*'s place in the body of knowledge began to slip, although the Aristotelian world picture still hung securely, if awry, in universities and theological seminaries. The slippage owed much to social factors associated, as in Islamic times, with the bureaucratic needs of newly centralizing states, and also with the discovery of new worlds on the Earth and in the heavens, which revealed that the ancients did not know even half of everything.

At the interface

While Alhazen was solving tough problems in geometrical optics, a polymath, Regimbaldus of Cologne, and a teacher, Rudolphus of Liège, tried to prove that the sum of the angles in a triangle is a straight angle. Not knowing Euclid, they had neither the proof nor the concept of proof, and ended by lining up the angles cut from triangles drawn on parchment. Although their mathematical simplicity was representative of their time and place, savants then existed in the Latin West who could handle hard computations easily. They were masters of *computus*, the science of the calculation of times of prayer, the dates of movable feasts, and the coordination of calendars. Their master, the Venerable Bede, a Northumbrian monk who died in 735, wrote on *physica* as well as *computus*, and dated events *anno domini* from the year now taken as the first in the Christian era.

Bede's *physica*—his understanding of the motions of the planets and the luminaries, of the rotations of the heavens, and the interconvertibility of the elements—came from Pliny's *Natural history* and the commentaries on the *Hexaemeron* by Saints Basil and Ambrose. From the same sources Bede learned that "the ocean tide follows the moon as if the luminary drew the waters behind it and then pushed them back" (thus the saints) and that, owing to the time required for celestial or meteorological events, such as thunder and lightning, to affect the Earth, a delay occurs between high tide and the Moon's meridianal passage (Pliny). But it was local knowledge that prompted Bede's important deduction, once acclaimed as the only original contribution to *physica* from the Latin West in 800 years: the difference in time between the meridian passage of the Moon and the succeeding high tide is a constant at any place. The riddle of the tides, which, according to legend, drove Aristotle to throw himself into the sea, thus acquired an additional complication.

Bede's knowledge made its way to the Continent via Alcuin of York, who served as Charlemagne's chief advisor in civilization. Alcuin revived the curriculum of the liberal arts, at least for the court clergy, and stimulated an interest in mathematics, at least of the recreational type. A little book attributed to him comprises 53 perennial problems, for example, river crossings under various restrictions (like the familiar puzzle of the missionaries and the cannibals), and divisions of property. His arithmetical games may have encouraged some students to go on to *computus*, on which he wrote a tract; but the "Carolingian Renaissance," which lasted about a century, did not establish a basis on which the likes of Regimbaldus and Rudolphus could build.

One of Alcuin's tasks was to write against a Christian heresy prospering in Umayyad Toledo. Some of the faithful there entertained unorthodox ideas about the person of Christ. They could not correct themselves as they could not read Latin, although they handled "the Arabic pomposity of language with the greatest erudition." Umayyad culture beguiled them further with gardens, libraries, and palaces, with scholars, artists, and musicians, and, for the technically minded, with improvements in metalworking, agriculture, waterworks, and trade. The straying sheep were saved from these attractions by the small Spanish kingdoms to the north, which, profiting from the usual divisions within Islamic states, captured Toledo in 1085. The Christian scholars who followed the Christian generals did not care for Arabic poetry. They wanted to learn *physica*, philosophy, astronomy, and mathematics, partly from curiosity, partly for practical application, and, perhaps mainly, to fortify themselves for arguments with learned Muslims. For, as one of the first scholars to work in Toledo wrote, "our civilization is infantile in comparison with theirs."

The earliest large-scale translation work in Spain mirrored the effort in Baghdad 400 years earlier. The archbishops of Toledo played the role of the early Abbasid patrons. Translators again invented technical terms via transliteration, bequeathing to English zenith, nadir, azimuth, almanac, algebra, Aldebaran and many other star names, alkali, alembic, elixir, alcohol, a number of things to eat (apricot, artichoke, sherbet), and indications of what made Islamic civilization prosper (bazar, tariff, admiral, arsenal, cotton, muslin, damask). And, as the scholars of al-Mamūn's House of Wisdom came from many parts of the Islamic world, so Westerners wanting wisdom came from every corner of Christendom. In addition to the Greek corpus, the scientific authorities awaiting their study included, to mention only names already encountered, Albumasar (Abū Ma'shar), Alhazen, Avicenna, Averroes, Avempace, Alfarabus, Alkindus, and Māshā' Allāh.

"A little learning is a dangerous thing | Drink deep or taste not the Pierian Spring." The first Latins who tried to apply Islamic astrological physics did not quaff deeply enough before deducing from an upcoming trigon change accompanied by a congregation of planets in Libra (an airy sign) that great winds from the East would dump deserts of sand on European habitations. Consulting their newly acquired authorities, Westerners feared the worst. A friendly Spanish Muslim astrologer wrote the Bishop of Toledo a letter of reassurance. The ignoramuses who made the predictions, "knowing nothing of the virtues of the heavenly bodies and the effects of the five planets and the two luminaries," had reasoned childishly. There was nothing to fear.

Alma mater

By coincidence, the West had just invented an institution in which to deepen its drinking. This was the *universitas* or guild organized by teachers accredited for their learning, chartered by the Pope, and supervised by the local bishop. (In a few cases, Bologna being the most prominent, students formed the corporation and hired the professors, see Figure 10.) Although the oldest university in the West, founded by modern consensus in 1088, Bologna dealt primarily in law; the *studium generale*, comprising at its fullest a preparatory arts faculty and professional schools in theology, medicine, and law, dates from a century or so later, concurrent with the foundations of the universities of Paris and Oxford. To add further dynamism and coincidence, the energetic new mendicant orders, the Dominicans and Franciscans, founded in 1216 and 1223, respectively, soon decided to establish chairs in the arts and theological faculties. Many of the greatest interpreters of Aristotelian *physica* active in mastering the learning of Greece and Islam were members of these vigorous orders. Thus, in significant contrast to Islam, the Roman Catholic Church embraced rationalistic *kalām* and produced a scholastic philosophy in which *physica* had an important place.

Most of the teachers or regent masters in arts were men engaged in the long process of earning a doctorate in theology. The degree they obtained, a bachelor's ("beginner's"), conferred the right to teach arts anywhere and thus, with other masters, create a new generation of artists, in a sort of academic apostolic succession. This *ius ubique docendi* reflected the universal claims of Christendom. Like their Islamic counterparts, Latin scholars had a common language and religion; unlike them, the Westerners had universities that gave institutional continuity and ordered competition in the study of general subjects deemed useful for theology and the lay professions of law and medicine.

Episcopal supervision, clerical professors, and a universal teaching credential did not create uniformity of curriculum or opinion. The arts masters favored Aristotle's philosophy, as it fell within their domain of knowledge attainable by reason alone. Theologians at first opposed it, not only for its novelty, but also on the sounder ground that it taught many things obnoxious to the Catholic faith. And so, according to its statutes of 1215, the University of Paris required candidates for the bachelor's degree in arts to have studied Aristotle's logic and ethics, but prohibited the public reading of his *physica* and metaphysics. After the *studium generale* in Toulouse (founded in 1229) advertised that it allowed these works, the Pope, Gregory IX, ordered the University of Paris to separate the useful material in them from the erroneous and scandalous.

Soon the arts masters at Paris were teaching an Averroist Aristotle, literally interpreting the texts and defending their presentations of erroneous and scandalous doctrines as fulfilling a commentator's duty to his author. This historical method of introducing naturalistic *physica*, which continued through the time of Galileo's teachers, did not prevail in thirteenth-century Paris. In 1277 the Bishop, Etienne Tempier, having received the recommendations of a committee of theologians, prohibited over 200 harmful propositions they claimed to have found in Aristotle. With this apparent abrogation of academic freedom, Tempier in fact liberated the Paris faculty from a slavish literal interpretation of the texts and opened the way to wider and wilder speculation. The University of Oxford, which, together with Paris, produced the most advanced *physica* of the later Middle Ages, adopted proscriptions similar to Tempier's.

The enduring correction of Aristotelian philosophy through revealed truth was the work of a cosmopolitan Saint, Thomas Aquinas, born in Italy, educated in Paris and Cologne under another saintly exponent of Aristotle, Albertus Magnus, and active as a Dominican master in Paris and Rome. Thomas agreed with Averroes that Aristotle had gone as far as possible on reason alone. To correct for revelation, Thomas replaced the unit Universal Mover by the triune Christian God and the Intelligences by angels, much as the Neoplatonists had done with the One and its emanations, and eliminated the naturalistic errors about Creation, the soul, the vacuum, the displacement of the world, and so on. However, he retained the Aristotelian structure, the quintessential celestial spheres, the four-ring circus of terrestrial elements, the meteorology, and all the paraphernalia of hylomorphism.

Dante took the resultant Christian world picture as the great theater for his journey through Earth to the Devil's umbilicus and thence through the antipodes, up Mount Purgatory, and past the planets to the divine presence beyond the sphere of the stars. On the way up, Dante stopped at the sphere of the Sun for a talk with the shade of St. Thomas, who appeared in a circle with 11 other sages. Bede was among them, furthest from Thomas; the nearest shades belonged to his teacher Albert and to Siger de Brabant, a champion of Averroist Aristotelianism and Thomas' unrelenting opponent in life. Their juxtaposition in heaven suggests that the claims of reason deserve a hearing even when in apparent conflict with revealed truth.

University physica

Since most early scholastic writers were theologians, they expounded *physica* either to clarify obscurities in philosophy or to apply it, after clarification, to doctrinal problems. Take charity. What prompts a philanthropist to give? Since in Aristotelian philosophy the question fell under the category of motion, theologians of charity could make use of concepts developed to analyze physical change. If a shove from the Holy Spirit causes the philanthropist to act, fundraisers should expect giving to slacken just as a ringing bell falls silent. Still, the ringing persists briefly and sometimes so does charity. What causes this persistence? The clapper transfers motion to the bell just as the Holy Spirit conveys a charitable impulse to the philanthropist. Thus the concept of conferred impetus, whether fleeting or enduring, settled into medieval *physica*. Applied to locomotion, it allowed arrows to fly without being pushed by the air and explained the acceleration of falling bodies as the sum of the impulses acquired in each minute or ell of descent.

The outlawing of Averroistic Aristotelianism at Paris and Oxford encouraged criticism of the fundamental concepts of form and essence. Does "dogginess" exist or is "dog" just the name we give to a group of animals for convenience? The second alternative, nominalism, had its major champion in William of Ockham. His approach to *physica* in the lectures he gave at Oxford as an inceptor (a bachelor of arts working at theology) irritated the Dominicans and some secular masters. They objected to his conclusions that essences or common natures do not exist, and that general propositions about groups or classes cannot be known to be true. Since the "Venerable Inceptor" (so called because he never finished his degree) held a strict voluntarism (nothing constrains God's action but the impossibility of compassing a contradiction) he taught that only experience, not a priori deduction, can establish what exists; and that in explaining the apparent relations among things, the fewest possible causes should be invoked ("Ockham's razor").

Nominalism favors description over explanation. Hence we find in fourteenth-century Oxford an innovation, introduced by Ockham's contemporary, Thomas Bradwardine, for discussing the grand question: how does the form of a substance change? Rather than vex themselves over the causes of change, Bradwardine and others at Oxford's Merton College distinguished between motions at constant speed and at uniform acceleration, and deduced that in locomotion the total distance traversed from rest in a given time at constant acceleration is half the distance that would be traversed at the maximum speed in the same time (the "Mertonian rule"). Although they gave increasingly complicated proofs of their rule, they made no direct applications of it, not even to free fall.

The most distinguished teacher of *physica* at the University of Paris in the fourteenth century was Jean Buridan, now known primarily for his ass—a donkey with a mind so logical and rigorous that it starved to death for lack of a sufficient reason to choose between two equally attractive bales of hay. Buridan was a nominalist concerned to carve a space for a living Aristotelian *physica* between the stultifications of Averroism and voluntarism. In a bold step he applied the concept of impetus to the celestial spheres. They would not need angels or Intelligences to inspire them if God gave them an initial push and the quintessence conserved the impetus. Buridan observed further that the Earth must jiggle. Owing to the unequal distribution of land and water, its center of gravity does not coincide with its center of figure. The ongoing processes of erosion and the self-movement of living creatures constantly shift the center of gravity, which in consequence hops about the center of the universe. Perhaps the Earth might even revolve around the center and allow the heavens to rest? Buridan did not build a systematic physics from these fragments. They should be appreciated, rather, as indications of the sorts of things that free thinking within an Aristotelian framework could produce.

Buridan's prize student, Nicole Oresme, invented a convenient and influential, though misleading, representation of the Mertonian rule. In it a horizontal axis or "longitude" represents space or time, its associated vertical axis or "latitude" the corresponding degree of an accidental quality like place, heat, or color (see Figure 11). In locomotion, the accident was place, the latitude

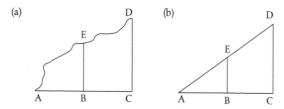

Fig. 11. Motion pictured. (a) AB the "longitude," usually time or space, BE the corresponding "latitude," or intensity of a changing form; ABCDE, the "total change" of the form over the entire longitude AC. (b) The case of motion at constant acceleration.

velocity, and the longitude indifferently space or time. The "latitude-of-forms" picture of motion at constant velocity is a rectangle, at constant acceleration a triangle or trapezoid.

One of Aristotle's arguments against the possibility of a void makes velocity proportional to the mover's push P and inversely proportional to the resistance R of the medium, from which the absurdity would follow that velocity in a vacuum would be infinite. To avoid this consequence and a finite velocity P/R when the resistance exceeds the push, Avempace had proposed an arithmetical rule, $v \propto (P - R)$. Bradwardine and, after him, Oresme, considered a complicated alternative that can be rendered anachronistically as $v \propto \log(P/R)$. It had no practical application apart from preserving Aristotle's argument against the void and making $v = 0$ when $P = R$.

The teaching that, according to later judges, made the fourteenth century the "Age of Buridan" occurred entirely in the Arts Faculty of Paris. A secular cleric rather than a regular, Buridan did not go on to theology but stayed in arts. In a development of first importance for all liberal studies, arts faculties strengthened against professional ones. A career in arts made long-term cultivation of special subjects possible and offered advancement with age, reputation, and administrative responsibilities. Whereas Bradwardine and Oresme pursued clerical careers (Bradwardine served for a month as Archbishop of Canterbury, Oresme for a few years as Bishop of Lisieux), Buridan rose to be Rector of the Arts Faculty and received several benefices to support his studies. But death levels distinctions, and the Black Plague indiscriminately ended the lives of the Rector, the Bishop, the Archbishop, and the Venerable Inceptor.

Mixed mathematics

Like the impetus theory of charity, statics (*scientia de ponderibus*) and optics (*perspectiva*) had a bearing on Christian religion. Although medieval statics could not calculate stresses in a cathedral, it provided some understanding of the effort needed to lift or slide weights into place and keep them there. For example, *physica* solved the problem of *gravitas in situm*, the tendency T of a body to slide down an inclined plane, which related to arches and domes. Thābit ibn Qurra and Avempace had made the tendency T equal to the weight of a body W diminished by the ratio of the plane's height h to its length L, $T = (h/L)W$. The most important of their Western followers, Jordanus Nemorarius, wrote in the early thirteenth century and was often copied and imitated. He combined the dynamic approach of the pseudo-Aristotelian *Mechanica*, and its virtual displacements of lever arms, with a geometrical analysis in the style of Archimedes;

his packaging of ancient and Arabic contributions to the science of weights served the West for two or three centuries. It was said that an expert in the *scientia de ponderibus* could weigh divine grace if he followed instructions with sufficient accuracy and faith.

The light of Creation, the Rainbow of the Covenant, and the filtered colors within Gothic cathedrals helped direct Christians to the study of optics. Robert Grosseteste, trained in Oxford and Paris, acted as director of studies in *physica* and mathematics to the Oxford Franciscans. Beginning at the beginning, Grosseteste imagined the light of Creation to have spread itself like a Neo-platonic emanation, creating space, the celestial spheres, and the rest of the cosmos. This *lux* (created light) is a general power manifested differently depending on where its *lumen* (propagating light) operates. Grosseteste's *lux* found its ideal medium in Roger Bacon, also educated in Oxford and Paris, where he taught Aristotle's *physica* and, in 1257, became a Franciscan. This proved to be a major error since the General of the Order, who became Saint Bonaventure, did not support studies not bearing directly on theology, and rejected two branches of *physica*, alchemy and astrology, that Bacon esteemed. Bacon's views continued to annoy his superiors, who jailed him in 1277, perhaps in connection with Tempier's prohibitions.

Bacon's development of Grosseteste's optics invoked the "multiplication of species," the spherical propagation of *lumen* from points of *lux*, which served also as a physical model of astrological influence. Bacon's *Perspectiva* incorporated Alhazen's optical principles, including image formation, except for intromission. Attempting to multiply his influence, Bacon designed a project to refute unbelievers. It required knowledge of languages, mechanics, optics, alchemy, and a *scientia experimentalis*, "wholly unknown to the general run of students." He wanted to check speculations about magnetic action, ways to extend life, and the projection of gold. Everlasting lamps and irresistible explosives also belonged to the unusual science of experiment, which Bacon did not practice much himself.

But he did make observations, and thereby put a number on Aristotle's factoid about the limited height of the rainbow. God placed a maximum of 42° on the altitude of the sign of the Covenant, a number that Bacon, who liked numerology, would have been pleased to know is the answer to the riddle of the universe according to the *Hitchhiker's guide to the galaxy*. In the first ever account of rainbow physics, God had written, "It shall come to pass, when I bring a cloud over the earth, that the bow shall be seen in the cloud ... And the bow shall be in the cloud." Subtle is the Lord! Did He mean that the bow exists in the cloud or that we only see it there? Bacon chose the second option and deduced that the observer creates the bow from rays undergoing reflection and refraction when passing through the droplets and air of variable density making up the cloud. Variations of this idea occur in influential works by other theologians: in a *Perspectiva* (largely a paraphrase of Alhazen) by Erazm Ciołek Witelo (a Polish theologian educated at Paris and Padua), and in a treatise *De iride* by Theodoric of Freiberg (a Dominican who studied and taught in Paris), which traced rainbow rays reverently and geometrically from the Sun through their reflections and refractions to the eye.

Fresh imports

From the Greeks

As the new universities assimilated their inheritance from Muslim writers, other material of equal importance became available from people fleeing a Byzantium wrecked by crusaders and threatened by other barbarians, and from investigation of material preserved in Italian churches and monasteries. Many Greek manuscripts thus emerged in Italy before the Eastern Roman Empire finally fell to the Turks in 1453. They included the works of Plato and the Neoplatonists, Greek versions of Ptolemy and Archimedes, and bits from the Stoics, Skeptics, and atomists. Just as a new institution, the university, coincidentally arose to receive the Arabic heritage, so a new social movement, Italian humanism, which started among learned secretaries serving the princes of new city states, identified itself with Greek language and literature.

The Greek *Almagest* that the most famous of the scholarly émigrés, Johannes Bessarion, later a Roman Catholic cardinal, carried to Italy, brought down the curtain on ancient astronomy. As we know, Muslim astronomers had pinpointed discrepancies between Ptolemy's predictions and observations, as well as the gap between his mathematics and his physics, long before 1453. However, as students tend to blame themselves for not understanding their teachers, so the astronomers of the West supposed that Ptolemy had all the answers; everything might be cleared up by consulting a Greek manuscript free from errors of translation and transcription. The outcome, secured mainly through the efforts of Bessarion's protégés Georg Peurbach and Peurbach's student Johannes Regiomontanus (University of Vienna), was decisive. The Greek *Almagest* did not resolve the problems of astronomy any better than versions made from the Arabic.

While the purified Ptolemy was dashing the hopes of humanist astronomers, other translators were recovering the truths preserved in Platonic treatises. Impressed by intellectual émigrés from Byzantium, Cosimo I de' Medici desired to resurrect the Platonic Academy in Florence. The great project of this revival was the translation not only of Plato but also of Plotinus and other Neoplatonic philosophers. Cosimo confided the task to Marsilio Ficino, who considered himself a Christian sage in the Platonic line. On Cosimo's orders, Ficino began not with Plato's dialogues but with the work of Hermes Trismegistus, a hodge-podge of Christian, Neoplatonic, and Gnostic thought put together in the third century CE. Most humanists regarded these writings as very ancient and Ficino, impressed by their apparent anticipation of Scripture, thought that Hermes was contemporary with Moses or, perhaps, Moses himself. Hermes taught among other things that celestial influences both good and bad can be captured in talismans as well as in people, and that, therefore, a *physicus* willing to experiment might add useful astrological powers to *physica*.

The Platonic Academy's Latinized Plato spread the Pythagorean doctrine that made mathematics the queen of *physica* rather than, as in the standard teaching, at best a handmaiden. In 1464, when Ficino had just started his translations, Regiomontanus promoted the Pythagorean vision in a lecture in Padua on Islamic astronomy and astrology. The lecture, decorated with humanist references, recommends the study of astrology as the highest, noblest, and hardest branch of mathematics, through which "we are brought near to Immortal God no less than through the other arts we are set apart from wild beasts." Regiomontanus drew from his ability to master the "angelic art" of astrology a confirmation of Hermes' dictum, "man is a *magnum miraculum*," perhaps as high on the scale of being as demons.

Among the astronomers encouraged by this insight was another Northern European perfecting himself as a humanist in Italy. Nicholas Copernicus, born the year Regiomontanus published Peurbach's epitome of Ptolemaic planetary theory (1473), studied at the University of Cracow and then, in 1496, went to Italy to learn law and medicine in preparation for an ecclesiastical career. He also studied Greek literature. His subsequent position as a cathedral canon in a region contested by Polish and German princes, and also by Prutenic Knights, left him little time to polish his version of proper astronomy. This turned on the objection pushed by Averroes, al-Tūsī, the Marāgha school, and Ibn al-Shātir: Ptolemy's use of improper devices like the equant point. Copernicus removed it using constructions that convert circular to rectilinear motion probably taken from Arabic astronomers ("al-Tūsī's couple"), although no persuasive path of transmission has been established. But the boldness and confidence that prompted Copernicus to go beyond removing the supposed blemishes of the equant to place the Sun at the world's center did not have a counterpart in Islamic astronomy.

The description of the apparent motions of the Sun, Moon, and stars is equivalent whether the Sun or the Earth stands still.

By making the Sun the center of motion, however, Copernicus could do what Ptolemy could not: calculate the distance from the Sun to a planet as a multiple of the radius of the Earth's orbit. His ordering of the planets and his consequent explanations of the limited elongations of Mercury and Venus, and the retrogradations of Mars, Jupiter, and Saturn, were the anchors of his claim to have made a system out of Ptolemy's bric-a-brac (see Figure 12). Copernicus' masterpiece, *De revolutionibus orbium coelestium*, appeared in 1543 with an unsigned preface by a Lutheran controversialist, Andreas Osiander, who saw the manuscript through the press. Anticipating objections from theologians, he rehearsed the usual arguments against taking mathematical constructions as truths of nature. Perhaps in consequence, the book did not excite theologians for 70 years. However, it caused an immediate ruckus among *physici*, who saw that it disagreed more profoundly with *physica* than Ptolemy's fictions.

Copernicus addressed some of the obvious objections, like the removal of the Earth from the universal center of heavy bodies, the lack of a power to move it in orbit, the fact that falling bodies do not land west of the vertical through their drop point, and so on. Some of his answers were physical (separated bits of planets fall to their centers, objects on a rotating planet participate in its motion) and some rhetorical, invoking a spectrum of ancient writers prized by humanists (Plutarch, Pythagoreans, Plato, Virgil, Hermes, Sophocles) who said things favorable to belief in a Sun-centered universe. Although some of Copernicus' answers pointed toward a viable new physics, they left much to do before an adequate replacement for standard *physica*, with a suitable role for mathematics, could even be sketched.

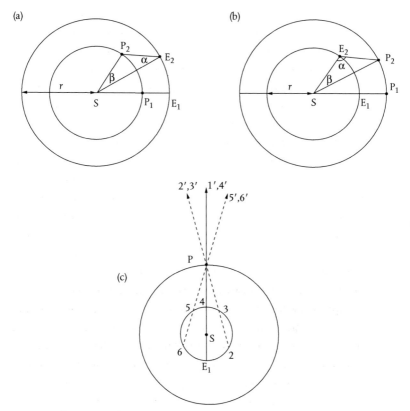

Fig. 12. Heliocentric advantages. Copernicus allows the computation of the solar distances of the planets P in terms of the Sun–Earth distance r. (a) The limited elongations of the "inferior planets" Mercury and Venus is a consequence of Earth's orbit encircling theirs; (b) the "superior planets" Mars, Jupiter, and Saturn can be seen anywhere because their orbits encircle Earth's; in both cases, α can be measured and β calculated from the known sidereal periods. (c) The superior planets appear to move backward, or "retrograde," against their usual cc motion as the Earth overtakes them, as between positions 2 and 4. For convenience, P is pictured as stationary.

While the cause of mathematics as natural philosophy went forward with the help of Plato's rediscovered eloquence, humanists brought forth another, greater threat to Aristotelian *physica* in the form of Latin poetry. This was the long-lost work of Lucretius, found in the early fifteenth century and printed for the first time in 1473. The notion that our world is but one of an infinite number composed for no purpose by innumerable atoms tumbling forever through infinite space was not wholesome doctrine for Christians. St. Jerome, who knew the poem, offered the monkish explanation of its perversity that its author went insane from an overdose of aphrodisiac. Nonetheless, the beauty of Lucretius' descriptions of the couplings of the atoms recommended his poem to all lovers of Latin, and helped to insinuate atomism in its various forms as an alternative to the *physica* of the schools, the mathematizing of the Platonists, and the Hermetic account of occult influences. Like Averroism, atomism became a dangerous subject during the confessional disputes of the sixteenth century, when the charge of heresy could be leveled to help win philosophical arguments, and Aristotelian *physica* became officially entangled with theology and the dangerous doctrine of the Eucharist.

From the real world

The tolerance of ambiguity in medieval thought—represented in great matters by compromises between reason and revelation, and in small ones by fuzziness in the concepts of impetus and the latitude of forms—began to give way to more exact forms of knowledge and action during the fifteenth and sixteenth centuries. The main causes of this hardening were the rise of powerful competing orders within the Roman Catholic Church; the Protestant reformation and the Catholic response; the discovery of the new world and its exploration; and the invention of printing. Religious controversy required clear definition of the points at issue among Catholic orders and Protestants sects. The Catholic Church replied not only by strengthening dogma and discipline through the decrees of the Council of Trent, but also by setting up a regime of thought police: the Holy Roman Inquisition (the Holy Office), the Congregation of the Index of Prohibited Books, and the Society of Jesus (the Jesuits).

On the secular side, the European discovery of America intensified the bureaucratic needs of the states competing in the new world. Managing the fleets, keeping track of the influx of precious metals, surveying the new possessions, and teaching and improving navigation, cartography, and geography opened new opportunities for applied mathematicians already in demand as architects, urban planners, and hydraulic engineers. Calculations, whether correct or not, depend on hard edges. Large collections of unfamiliar material brought back from exploratory voyages, and reports of the strange behavior of peoples who did not have the happiness of being European, increased the demand for minds able and energetic in denying and distinguishing. And, to complete this hasty inventory, printing, by replacing costly and imprecise manuscript copying with more-or-less exact reproductions, gave scholars relatively easy access to standard texts. Standardization and definition did not resolve disputes, but made it more likely that the parties to them argued about the same things.

Many consequences for *physica* of the hardening of the edges and the standardization of texts may be discovered in Galileo's work. When professor of mathematics at the University of Padua, from 1592 to 1610, he took in private students of aristocratic rank who came to learn the use of surveying and measuring instruments. His expertise in designing practical devices enabled him to transform a spyglass of little magnification into a 30-power telescope in a few months. The printing press helped "the new Columbus" to spread news of his implausible telescopic discoveries—mountains on the Moon, satellites around Jupiter, stars in the Milky Way—as it had given him access to the books of Euclid, Archimedes, Ptolemy, and Copernicus, from which he drew inspiration. When he interpreted his discoveries in ways that conflicted with the views of the Pope and the Inquisition, the hard-edged institutions of the Catholic Counter-Reformation came into play, and condemned the Copernican world picture. Galileo's conclusions catapulted astronomy into the contested arena of *physica* and himself from a professorship of mathematics at the tolerant Venetian University of Padua to the dangerous height of "Philosopher and Mathematician to the Grand Duke of Tuscany" in the bigoted Medici court in Florence.

Galileo's friend Pope Urban VIII was a cultured Florentine and man of the world. Yet he believed on grounds stronger than Plato's that the unaided human mind cannot arrive at secure truths about the natural world. Urban had followed the logical track of God's Omnipotence and Freedom to an extreme voluntarism and the conclusion that, however well our theories fit the facts, God could have contrived to bring about the same phenomena in countless other ways. Galileo countered that his business was not to imagine how God might have done things, but to discover the method He chose. His spirited defense of Copernican theory, made in his *Dialogue on the two chief world systems* (1632) after the Holy Office had found heliocentrism contrary to Scripture and "philosophically absurd" in 1616, precipitated his trial, condemnation, humiliation, and perpetual house arrest.

It was not this suffering that made Galileo a hero to later physicists, however, but his attack on *physica* under the banner of mathematics. His target at first was the nub of Aristotelian philosophy, the doctrine of motion. When a young professor at the University of Pisa, he replaced the supposedly Aristotelian formula, $v \propto P/R$, with Avempace's rule, interpreting $P - R$ in Archimedean terms as the difference between the specific gravities of the falling body and the resisting medium. If Galileo dropped weights of different materials from the Leaning Tower as the legend started by his last pupil, Vincenzio Viviani, has it, he would not have expected them to descend at the same speed.

Galileo came to greater enlightenment by abandoning the search for causes. He gave up on Archimedes as he had on Aristotle and entered into a free fall with no goal in sight other than a mathematical description of the journey. He found and confirmed by some mixture of experiment and conjecture that (as he first expressed it in 1604) bodies descending freely from rest cover, in successive intervals of time, spaces proportional to the odd numbers beginning with one. Representing his rule by a graph identical to the medieval picture of the latitude of forms, with instantaneous velocity as the horizontal axis and space traversed as the area, he had, as he wrote it, $\Delta ABE:\Delta ACD = AB^2:AC^2$ (Figure 11b, rotated 90°). Since he intended the graph not as a qualitative illustration but as a representation of mathematical relationships, he had to decide what the latitudes and the triangles stood for. Initially he fell for the obvious choice, taking the vertical axis to be the distance of fall. Eventually he realized that the choice led to impossibilities. Taking then the vertical axis as time, he had a successful plot, at the major expense, however, of the counterintuitive identification of a line with time and an area with distance. With this sacrifice or advance he achieved the now celebrated, but then not obviously useful, result that distance traversed under constant acceleration from rest is proportional to the square of the elapsed time.

Galileo's kinematical method, ignoring forces and causes, was perfectly adapted to the contest over world systems and probably owed something to it. The traditional circles and epicycles, which turned without any specified force, and which Copernicus also employed, are kinematical devices. As everyone knowledgeable knew, without a physics (and a theology!) to determine the world's center, the circles might be referred to the Sun or the Moon and the appearances be saved. Hence neither of the chief world systems, nor their chief competitor—the system worked out by the most exact of observers, Tycho Brahe, in which the planets go around the Sun while the Sun circles the Earth—could be affirmed to be true on the basis of astronomical observations alone. Galileo tried to go further, and to reduce gravitational fall and the effect of the Earth's rotation to purely kinematical and geometrical relations, and so defeat the standard physical arguments against the Earth's diurnal rotation. He also devised a kinematical argument to show that only if the Earth moved as Copernicus said it did would there be diurnal, monthly, and annual tides, and published his theory as the provocative conclusion to his *Dialogue*. To counter the impression that he believed in the condemned Copernican system, Galileo saved the day but not himself by having the dunce of the *Dialogue* undercut all world systems with Urban's voluntarist epistemology. The Pope never forgave Galileo for trashing his powerful philosophy of science, with which he had hoped to answer all challenges to Scripture and papal dicta that *physica* could mount.

Galileo situated his *Dialogue* in a place fit for a discussion of liberal arts—a palace on the Grand Canal in Venice. He chose the Venetian Arsenal, a shipbuilding complex, for his *Discourses on two new sciences* (1638), which treated the strength of beams and exterior ballistics. Although his geometrical analyses of these problems were of little practical value, they indicated what a quantitative applied physics might look like; and his technique of compounding a constant velocity imparted by the cannon with an accelerated drop caused by gravity, from which he derived the ideal parabolic trajectory of missiles, became fundamental to analytical mechanics.

In moving his discussants from palace to workshop, Galileo made common cause with a few physicians and other university-trained men who took a sustained interest in the physical principles of practical devices. One of the doctors, Georg Agricola—a learned humanist who practiced in mining regions—illustrated extractive and metallurgical techniques so as almost to thrust the concepts of force, momentum, and pressure into the reader's mind (see Figure 13). The other doctor, William Gilbert—an English physician fascinated by the magnetic compass that enabled the trade and piracy of London—studied lodestones intensively in what was the first sustained, systematic experimental investigation in the long history of *physica* (*De magnete*, 1600). Gilbert invented instruments to measure the dip and declination of a compass needle moved around a spherical lodestone, which performed like a little earth, a *terrella*, magnetically speaking. His method of arming magnets to increase their power instructed Galileo, and his premise that the Earth was a big magnet inspired Kepler. In vituperation of his opponents, "intoxicated, crazy, puffed-up...lettered clowns," and his evaluation of his accomplishments, "novel, unheard-of, and opposed to everybody's opinions," he almost outdid them both.

A final example, Simon Stevin, made his living as an engineer and state administrator after a late matriculation, in 1583 at the age of 35, at the University of Leyden. The slow start brought him into contact with a fellow student, Maurice of Nassau, who would soon become the head of the United Netherlands. Stevin served Maurice as tutor, military engineer, and quartermaster, and worked with Dutch-reading applied mathematicians as an authority on statics, astronomy, and fortification. Although, unlike Galileo, Stevin made practical devices that worked, they took up many of the same theoretical problems and arrived at similar results, for example, the composition of velocities and the strength of gravity on planes of different slopes. Stevin declared for Copernicus publicly in 1608, before Galileo did, and proposed a method to find the longitude at sea using

Fig. 13. Applied physics around 1550. A series of pumps removing water from a mine. The series overcame nature's limit of a 30-foot rise for a single pump. From Georg Agricola's *De re metallica* (1556), a coffee-table book describing rough technology in polite Latin.

magnetic data no more practical than Galileo's scheme using eclipses of Jupiter's moons. In short, Stevin wrote on almost the entire canon of mixed mathematics, and would have been more influential in his time, and better known in ours, had he not written it in Dutch.

Fig. 14. Royal Academicians. Louis XIV's main minister, Jean-Baptiste Colbert, presents the Académie royale des sciences to its master. The unfinished Royal Observatory (completed in 1671) appears in the background.

4

A SECOND CREATION

---•◦•---

Revolution or integration?

"New philosophy calls all in doubt | The element of fire is quite put out | The earth's lost and no man's wit | Can well direct him where to look for it." John Donne thus recorded the tremors in a world of learning shocked by the discoveries of the Columbus of the heavens and of the incompetence of all ancient systems of thought, apart from the skeptical, to accommodate modern knowledge. For a time the philosophical fashion was skepticism, which drew strength and respectability from the recovered skeptics' handbook of Sextus Empiricus (around 200 CE), and made common cause with its emotional opposite, the voluntarism of Ockham and Urban.

The specter of skepticism, of throwing all philosophy into doubt, inspired three important schemes for placing *physica* on firm foundations or limiting its claims to what could be defined as secure natural knowledge. By the middle of the eighteenth century, Newton's scheme had won out over those of his rivals,

Descartes and Leibniz, although only by dropping epistemo-logical scruples that had weighed with all of them. This renun-ciation reduced their rigorous logic to the easy-going "reason" of the Enlightenment, and replaced their specifications of God's place in His creation with a vague assimilation of God with Nature. Unscrupulous or enlightened Newtonianism prepared the way for the invention of physics and its first standard model by overcoming the dichotomies between mixed mathematics and *physica*, and between celestial and terrestrial mechanics.

Manifestoes and cadres

The most promising approach for early seventeenth-century sappers of Aristotelian *physica* was to build a celestial mechanics for a Copernican universe. The appearance of an auspicious comet in 1577—which, according to parallax measurements by Tycho and others, must have sailed through the planetary regions—enforced the impression that the heavens must be "liquid" rather than crystalline. The supernovas of 1572 and 1604, likewise located beyond the Moon, disclosed that quintes-sence was liable to change. Tycho's sometime assistant and successor as mathematician to the Holy Roman Emperor, Johannes Kepler, not only freed the planets from their 2,000-year crystalline imprisonment, he also broke with all authority—ancient and modern—by holding that the planets employed their freedom describing elliptical orbits with the Sun at a focus. He kept them in these orbits by a far-fetched combination of Gilbert's magnetic philosophy and Aristotelian dynamics. From Gilbert he concluded that the Sun contains a magnetic monopole and the planets magnetic dipoles with directions fixed in space. For an Aristotelian constant push he relied on rays from the Sun, which he supposed to turn on its axis, sweeping the planets around in circles that the magnetic forces distort into ellipses. Kepler's solar lighthouse anticipated Galileo's discovery of the Sun's axial rotation by several years.

A younger contemporary of Gilbert's at the court of Elizabeth I, Francis Bacon, who rose to become the chief lawyer in the land, indicted the school philosophy on several capital charges. He did not attempt a new celestial mechanics in the style of Gilbert or Kepler, but proposed a new method of inquiry, a new purpose for *physica*, and a new institution for pursuing it. The method was to collect all the information available on a given topic, extend and correct it by experiment, winnow and systematize the results, and, by "crucial experiments," decide between different theoretical alternatives. The new purpose was "the improvement of man's estate;" "knowledge is power," Bacon said, in one of his enduring phrases, and *physica* without practical applications is no knowledge at all. As for the institution, Bacon proposed to Elizabeth's successor, James I, that he equip a house of Solomons with all the books, instruments, and materials needed to devise a replacement for Aristotle. This would not be the leisured amusement of free men, nor the narrow speculations of monks, nor even the bookish science of university professors, but a systematic aristocratic experimental enquiry directed to the exploitation of nature (see Figure 14).

Although in the mid-seventeenth century the Jesuits, the schoolmasters of Catholic Europe, still insisted on Aristotelian *physica* as fundamental to their philosophy and theology, most knowledgeable people knew that it could not last much longer as the blueprint of the physical world. Atomism offered a possible replacement. Galileo had endorsed it in his anti-Jesuit *Saggiatore* (1623) and Pierre Gassendi, a priest and professor in Paris, Christianized Epicurus as a paladin against Aristotelians and skeptics. Nonetheless, Gassendi ended up near skepticism himself by allowing God the freedom to alter the order of nature. Eliminating skepticism and voluntarism to gain space for physics was the providential mission of René Descartes, whose famous bout of doubt ended, as he reported in his *Discourse on method* (1637), with the assurance that God would not deceive him about any conclusion he apprehended with the same clarity and distinction as the inference "I think, therefore I am." This principle required God to have made necessary what Descartes conceived as such, and to stick to His original scheme; if He had thought it good enough once, it would derogate from His wisdom to change His mind merely to show His power. Descartes backed up these gratuitous assertions with spectacular acquisitions he claimed to have made by his method, which he exhibited in *Essays* appended to his *Discourse*.

One *Essay* presents his invention of analytic geometry—a proof, he said, that the ancients did not know everything about mathematics. A second, on optics, shows how to improve telescope lenses and how the lenses and muscles of the eye work. The third essay, on meteors, provides a corpuscular mechanism for most of the physical phenomena Aristotle treated in his *Meteorologica* and Pliny crammed into his *Natural history*. It also offers a singular result, obtained using the law of refraction, $\sin i = n \sin r$, now known as Snell's law after Willebrord Snell (Leyden), whose papers Descartes probably saw during his many years' residence in Holland. Applying Snell's law to raindrops, he calculated, from the known value of the index of refraction of water n, that the maximum height of the primary bow must be 42°, the value observed by Roger Bacon almost four centuries earlier.

Descartes' *Principia philosophiae* of 1644 extends the corpuscular approach to the full spectrum of *physica*. It begins with laws of motion expressed mathematically, which, in principle, regulate the intercourse of the differently shaped bits making up the physical world. These bits are not atoms, as Descartes had a horror of the void as great as Aristotle's, and, despite his claims, they do not obey his laws of motion and possess something like mass as well as mobility and extension. Still, to make a universe out of mass, motion, shape, and the impossibility of a vacuum would challenge a Demiurge. Descartes met it, like another Timaeus, with a creation myth, which derives the present state of the universe from an original undifferentiated space-matter activated by God's command: "let there be motion."

The myth describes how the heavenly bodies, each of which once sat as a sun at the center of its own vortex, differentiated by being swept into their neighbors' whirlpools. Earth thus captured the Moon before both were torn into the Sun's maelstrom. What the Earth–Moon system retained of its original vortex causes the fall of bodies, the tides, and the magnetic phenomena detailed by Gilbert (see Figure 15). Particles of various shapes, and bodies with special pores, account for all the activity in the visible universe including the machinery of the body and the operation of the senses. The only substance in the world besides the universal space-matter is mind, which, by a mysterious

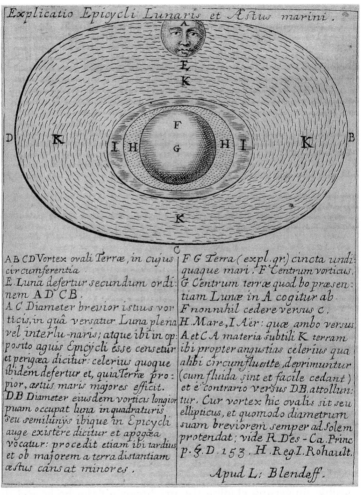

Explicatio Epicycli Lunaris et Æstus marini.

A B C D Vortex ovali Terræ, in cujus circumferentia.

E Luna defertur secundum ordinem A D C B.

A C Diameter brevior istius vorticis, in quâ versatur Luna plena vel interlu-naris; atque ibi in opposito aguis Epicycli esse censetur et perigæa dicitur: celerius quoque ibidem defertur et, quiaTerræ propior, æstus maris majores efficit.

D B Diameter eiusdem vorticis longior quam occupat luna inquadraturis seu semiluniys ibique in Epicycli auge existere dicitur et apogæa vocatur: procedit etiam ibi tardius et ob majorem a terra distantiam æstus causat minores.

F G Terra (expl. qr.) cincta undiquaque mari. F Centrum vorticis.

G Centrum terræ quod bo præsentiam Lunæ in A cogitur ab F nonnihil cedere versus C.

H Mare, I Aër: quæ ambo versus A et C A materia subtili K terram ibi propter angustias celerius qua alibi circumfluente, deprimuntur (cum fluida sint et facile cedant) et è contrario versus DB attolluntur. Cur vortex hic ovalis sit seu ellipticus, et quomodo diametrum suam breviorem semper ad Solem protendat; vide R Des - Ca Princ p. 9 D 153. H Reg I Rohault.

Apud L: Blendeff.

Fig. 15. The Earth–Moon eddy in the solar vortex. The remnant of Earth's primeval vortex ABCD carries the Moon, and, flowing more quickly in the narrow passage between the two bodies, creates the tides.

connection to the body at the pineal gland, supplies all the color, sensation, and drama of life.

Descartes did not deliver his wide and whimsical world picture as true in all details. The creation myth does not agree with what Moses tells us of God's methods, and Descartes could not be certain which of the many mechanisms he could conceive God had chosen. His guesses merely indicated that his system could be made to work quantitatively. Whereas Galileo had quantified a small portion of the science of motion in a manner Descartes criticized as spotty, unsystematic, and indulgent, he, Descartes, had demonstrated that something like a mathematical physics was possible and even useful. And, in contrast to Aristotelian philosophy—which only trained philosophers who knew Latin might think they understood—Descartes' spare system lent itself to development using intuitive concepts available to everyone who could think and read French.

The earliest cadre of Cartesians consisted of Dutch physicians attracted by the project of basing medicine on mechanical accounts of the body's functions and repelled by old-fashioned medical establishments opposed to new ideas. In France, well-placed lawyers, at odds with a regime that rated ancient privilege above existing talent, attended lectures by Descartes' disciples. One of their gatherings turned into the Académie royale des sciences in Paris (founded 1666). Also, many of the moving spirits behind the Royal Society of London (founded 1662), for example, Robert Boyle, had flirted with Cartesian philosophy. And so the creation of the first enduring academies of science coincided with the promotion of the first natural philosophy comprehensive enough to replace Aristotle's. Like the simultaneous appearance of Aristotle's *physica* and the universities in which to cultivate it some five centuries earlier, the incorporation of corpuscular philosophy in academies of science made a revolution.

The Scientific Revolution had an effective engine of war that Descartes had not thought of. It was the air pump, which debuted in 1654 in Regensburg, where its inventor, Otto von Guericke, Mayor of Magdeburg, brought it to amuse his fellow delegates to meetings called to clean up the mess left by the Thirty Years War (see Figure 16). The extraordinary power of nothing came increasingly under the control of experimenters, as Boyle, his sometime assistant Robert Hooke, and Christiaan Huygens (the leading member of the Paris Academy of Sciences) began improving Guericke's instrument. Spaces devoid of air were the first reliably reproducible artificial environments in which to "twist the lion's tail," that is, in Bacon's image, to experiment on nature.

The earliest such experiments, performed in the void space above the liquid in a barometer tube, date from the early 1640s when a mathematician in Rome, Gasparo Berti, using a barrel, water, and a lead pipe, investigated why ordinary pumps could not lift water higher than 30 feet. Galileo's prize student, Evangelista Torricelli, who succeeded him as Grand Ducal Mathematician, replaced the barrel, lead, and water with a small basin, glass, and mercury, and so created a compact instrument for measuring heights, tracking the weather, and testing the nature of apparently empty spaces. Such tests preoccupied the Accademia del Cimento, a short-lived group of Galileo's followers supported by the Medici between 1657 and 1667.

Fig. 16. Diversion from politics. One of the demonstrations with which Otto von Guericke entertained delegates at Regensburg in 1654. The lid of the heavy cylinder cannot be removed by gravity or political maneuvering after the cylinder is evacuated.

They found that a void can transmit light and magnetism, but not sound and electricity, which Gilbert had proved acted by a mechanism different from magnetism's. Evidently nothing has a capacity to discriminate.

Cartesians tried to redefine vacuum, but having little to work with found it hard going. It became harder after Blaise Pascal had explained that the barometer column stands because the atmosphere exerts pressure on the mercury in the basin and none in the empty space at the top of the tube. Boyle and Hooke put some numbers on the pressure by sealing a sample of air in a U-tube closed at one end and pouring mercury in at the other. In 1662, in a rare early example of quantification in *physica*, Boyle published their finding that the air's spring is as its density (pressure $\propto$ volume^{-1}). Hooke found a similar relationship for a real spring: force is as extension, *ut tensio sic vis*, or, as originally formulated as an anagram to secure priority and monopoly, "ceiiinossssttuv."

As revolutions should, the Cartesian one immediately stirred up opposition. The curators of Dutch universities made common cause with Jesuit generals in prohibiting the teaching of any philosophy but Aristotle's. The Archbishop of Paris shut down public lectures on Descartes' physics, the Holy Office condemned his philosophy, the Index prohibited his works, and the Papal and Spanish authorities who ran Naples jailed doctors and lawyers who defended his ideas. The smoke cleared around 1700. By then the enlightened Cartesian had become the prototypical academic in Paris. The Oratorian Nicolas Malebranche was exemplary. When fresh from his scholastic training, he could not peruse Descartes without palpitations of the heart. Gradually he grew stronger and advanced from Aristotelian darkness to the light of mathematics, the harmony of sound physics, and a chair at the Academy. Malebranche's sound *physica* eased some of the problems of Cartesianism with the uneconomical doctrine of occasionalism, which considers pushes (the only source of physical change in Descartes' desert world) as occasions for God's continuous and ubiquitous action on His creation.

Mathematics or physics?

When Newton entered Cambridge as an undergraduate in 1661, the advanced thinkers there affected the philosophy of the "great secretary of nature," Descartes. Much of Newton's physics developed from an effort to grasp, demolish, and supersede Descartes. He began the work in 1665–6 while rusticating at his mother's farm to escape a plague raging in Cambridge. There he encountered the legendary apple. Two decades later, while professor of mathematics at Cambridge, he put together the lesson from the apple with constellations of problems found in Descartes' *Principia*. The title Newton gave his masterwork reflects the inspiration he obtained from his predecessor's work and the main fault he found in it. Whereas Descartes' *Principia philosophiae* dealt, like Aristotle's *physica*, with the foundation of all philosophy, Newton's *Principia mathematica philosophiae naturalis* restricted his to what he regarded as the crucial elements of a mathematical physics.

Newton's first fruitful engagement with Descartes concerned the improvement of telescopes. It led to an experimental study of prismatic colors, which for its ingenuity, method, and care set a new standard in physical science. Newton modestly described its main result as "the oddest if not the most considerable detection wch hath hitherto beene made in the operations of nature" when transmitting it to the Royal Society of London, then, in 1672, only a decade old. He had found that sunlight, which all students of optics had treated as pure and homogeneous, consisted of rays of different colors. Ingenious crucial experiments showed that the rays derived from sunlight could not be further divided. To each ray corresponded a different and definite index of refraction.

Newton added to this kinematics of color a conjectural dynamics in which the rays of light consist of particles shot from incandescent bodies. Their release sends fast pulses through a space-filling springy ether with which he sometimes replaced or filled the void. A ray encountering a transparent or translucent body crosses its surface or rebounds from it according to the phase of the ether vibration there. If toward (or away from) the denser medium, the ray has a "fit" of easy transmission (or reflection). Neither Newton's chromodynamics nor "the oddest detection" it explained pleased *physici* wedded to the hoary concept of color as impure light. Among the loudest of Newton's critics was Hooke, who had moved from assistant to Boyle to curator of experiments at the Royal Society, and whose expertise the new theory directly challenged. While Newton vegetated on the farm, Hooke published his magnificent *Micrographia*, which, besides memorable detailed illustrations of a flea and a louse and other personal items seen through his microscope, discussed color as modified white light.

The controversy with Hooke and others caused Newton, who preferred to dictate than to argue, to lay aside his work on light and color until Hooke died in 1703. Newton then published a big book on *Opticks* (1704) and accepted the presidency of the Royal Society, which he ruled from 1703 until his death in 1727. The

Opticks served as an up-to-date guide to the curators of experiments at the society since its successive editions added a growing number of "queries" ranging freely through *physica*. The conjectural functions of ether (or several ethers) increased to cover gravity, magnetism, and electrical attraction. In parallel with these conjectures, Newton suggested others that allowed the particles of bodies to interact directly via forces acting through space. Whatever their plausibility, his conjectures, like Descartes' vortices, applied the same physical principles not only to celestial and terrestrial phenomena, but also to the unseen world. This presumption—that principles adequate to explain the visible universe apply universally—which Newton stated explicitly as a rule of philosophizing, guided physical theory until the twentieth century showed its untenability.

Newton's glory was to fulfill, in his *Principia* of 1687, Galileo's hope of geometrizing gravitation. The ingredients of his solution were Descartes' first law of motion (the principle of inertia), Galileo's rules of free fall and the composition of velocities, and Kepler's prescription that a planet moves along its orbit so that the line joining it to the Sun sweeps out equal areas in the ellipse in equal times. Using the inertial principle and the composition of velocities, Newton showed that Kepler's prescription followed from the assumption that a planet accelerates toward the Sun along the line joining their centers. Since the prescription allows the substitution of an area for a time, Newton could reduce the problem of the magnitude of gravitational acceleration, assumed to decline with distance, to a problem in geometry. He found that if the attractive center is at a focus of the ellipse, the strength

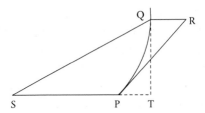

Fig. 17. Planetary motion geometrized. Suppose the planet moves from P to a close point Q in the very short time t at a constant acceleration a toward the fixed center S. If instead it escaped from S's control at P, in time t it would have moved along the tangent to, say, R. By Galileo's rule, $a = 2QR/t^2$; according to Kepler's second law, area SPQ $\approx$ SP $\times$ QT/2 is proportional to t; consequently $a \propto QR/(SP \times QT)^2$. It requires only geometry and cleverness to show that if PQ is part of an ellipse with S at a focus, $a \propto 1/SP^2$.

of the acceleration declines inversely as the square of the distance between Sun and planet (see Figure 17).

According to the apple story, the idea of universal gravitation hit Newton when he perceived that the Moon—being, as Galileo had argued, nothing but a big rock—should fall toward Earth's center as an apple does. Assuming that the Earth acts on both apple and Moon as if its entire accelerative power were located at its center (a theorem Newton later proved as a consequence of the geometry of a homogeneous gravitating sphere), he estimated the acceleration a of the Moon and compared it with the amount required in his theory to keep it in circular orbit: $a_M : a_A = d_A{}^2 : d_M{}^2$, where A, M designate apple and Moon, and d distance from the Earth's center. Taking $d_M \approx 60r$ (r being Earth's radius) and $a_A \approx 32$ ft/sec^2, $a_M = 32/3,600$ ft/sec^2. But also, to keep the Moon in a circular orbit, $a_M = v^2/d_M$, where v, the Moon's linear velocity, is $2\pi d_M/\tau_M$, $\tau_M = 27.5$ days being the period of the Moon's revolution around the Earth. On this reckoning, $a_M \approx 33/3,600$ ft/sec^2. It followed that if Newton could throw an apple hard enough he could turn it into a moon.

As they circle the Sun in their mutual embrace, Earth and Moon revolve around their common center of gravity (see Figure 18). At the Earth's center, the Moon's attraction just balances the tendency to fly off on a tangent. On the side of the Earth nearest the Moon, therefore, the attraction is stronger than needed to counter centrifugal force; on the far side, weaker; and so water is pulled up under the Moon and thrown out opposite it. Thus the two diurnal tides. The additional pull of the Sun accounts for springs and neaps. Newton thus quantified several of the enduring problems of *physica* that Descartes had bundled together: the nature of light and color; the cause of the tides and the principles of planetary motion; and the size and shape of the Earth. The shape turned out to be only approximately spherical. The gravitational theory, applied to a spinning blob of molten rock (the Earth in Newton's creation myth), brought forth an oblate spheroid, a pumpkin, several miles longer in its equatorial than in its polar axis.

Unfortunately, the *Principia* was not physics. Or so said the reviewer in the *Journal des sçavans*, which inclined toward the philosophy of Descartes. As Newton acknowledged, he had not assigned a physical cause to gravitation, but had recourse to a mathematical fiction, an immediate action at a distance. From a Cartesian point of view, he had retrograded. Either he had reintroduced occult causes (the hidden causes of manifest effects) with which the schools avoided explanations or, taking at face value that he wrote merely descriptively of observables

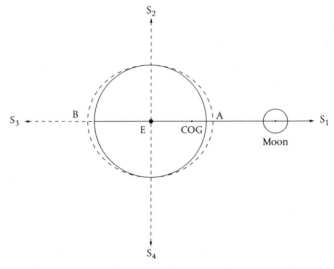

Fig. 18. Newtonian tides. Newton's gravitational theory requires the Earth and Moon to saunter around their common center of gravity (CoG). Since centrifugal and gravitational forces balance at Earth's center E, the latter exceeds the former under the Moon and the former the latter opposite the Moon. The dashed oval represents the strength of the net vertical force, not a bulge in the sea; the driver of the lunar tides is the component of the net force parallel to the water surface. At S_1 (new moon) and S_3 (full moon) spring tides occur; at half moon (S_2 and S_4), neap tides.

like accelerations, he had used the old dodge of separating mathematics from physics when the going got tough.

Newton replied to this criticism in an eloquent "General Scholium" added to the second edition of the *Principia* (1713). He declared that the universe is not the work of an absentee mechanic, but proceeded "from the counsel and dominion of an intelligent and powerful Being ... eternal, infinite, absolutely perfect." How the Being contrives gravity, His servant Newton did not know, and "I frame no hypotheses, whether physical or metaphysical, whether of occult qualities or mechanical." Exact description sufficed. "[T]o us it is enough that gravity does really exist, and act according to the laws that [I] have explained, and abundantly serves to account for all the motions of the celestial bodies, and of our sea." Thus Newton completed one full turn of the helix of scientific advance. His gravity resembled Aristotle's, his omnipresent God the Stoics' *pneuma*, and his planetary theory, with its preference for mathematics over physics, Ptolemy's artificial circles. In his reversion to these elements of the *ancien régime*, in his system of laws and his efforts to impose them, and in his imperial presidency of the Royal Society, Newton was the Napoleon of the Scientific Revolution.

Newton aimed the carefully crafted epistemology of the *Principia* against skeptics and *physici* who objected to the methods he claimed to be free from hypotheses. For people like himself who worried that his system did not offer enough employment for God, he assigned the Creator the task of pushing the planets now and then to insure they did not fall into the Sun. This tinkering seemed ridiculous to the librarian and historiographer of the Elector of Hannover, Leibniz, who had fashioned his own response to skeptics and providentialists. His God, like Descartes', did not behave arbitrarily; but whereas Descartes allowed God a choice of laws and initial conditions, Leibniz observed that His character and reputation required that He make the best of all possible worlds. Although we are not acquainted with all the reasons that make our world the best, we can be sure that sloppy workmanship was not one.

This conclusion gave rise to correspondence between Leibniz and Newton's spokesman Samuel Clarke that may be regarded as the high point in the ancient effort to define the divine implications of *physica*. Leibniz did not counter Newton's Tinkerer with an Absentee Creator, but with an Occasionalist. Unable to give a logical account of the collisions of atoms or corpuscles, whether elastic or inelastic, Leibniz decided that they do not act on one another. They are "monads," separate worlds, whose collisions follow the laws of billiards owing to a "pre-established harmony" between their motions and God's ordinances.

The invention of physics

The practice of demonstrating the new *physica* to a fee-paying public via experiments had a vogue in Cartesian circles after 1670. The notable figure here, Jacques Rohault, described his offerings in a text that Clarke thought proper to translate into Latin with prophylactic Newtonian notes. Hooke's successors at the Royal Society during Newton's presidency, Francis Hauksbee and the Huguenot J.T. Desaguliers, freelanced as public lecturers. Perhaps the most successful such performer was Jean-Antoine Nollet, a peasant's son who rose to be tutor to the royal children, lecturer at the royal artillery school, and a member of the Paris Academy of Sciences (see Figure 19). Beginning around 1720, experimental demonstrations of Newtonian physics entered classrooms, notably those of W.J. 's Gravesande and his successor Petrus van Musschenbroek in Leyden. Their textbooks defined the scope of experimental natural philosophy (or *physique expérimentale*) around 1750.

N. le Sueur Invenit   R. Brunet fecit.

Fig. 19. Fashionable lecturing around 1750. The lecturer, the abbé Nollet, whose clientele included the royal family, is charging a human condenser.

It was far narrower than traditional *physica*. To attract and please a sizable and discriminating public, the demonstrations had to be clear, clean, and easily visible. These requirements, through which the wider society helped define the *content* of physics, limited the subject to mechanical principles and model machines, pneumatics, optics, capillary and other hydrostatic phenomena, heat, magnetism, electricity, and astronomy as presented through the orrery (a mechanical solar system invented around 1704), armillary, and globes. After the accidental invention of the Leyden jar or condenser in 1745, electricity became the main attraction in this repertoire and Nollet its main expositor. His dominance did not last long. At mid-century, after exposure to sparks and glows by a public lecturer in Philadelphia, Benjamin Franklin invented the system of positive and negative electricity. When interpreted as centers of Newtonian forces, Franklin's electrical charges drove Nollet's Cartesian theory of electrical matter from the field. Franklin had a genius for devising electrical demonstrations, the most famous of which, flying a kite to catch a thunderbolt, he probably had the good sense not to try.

Electricity was also the bellwether in the quantification of physics. Charles Augustin Coulomb, an engineer trained at the artillery school in which Nollet taught and eventually a Parisian academician, devised an instrument that balanced the force between two electrified pith balls against the measurable torsion of a wire. Although few people if any have ever been able to reproduce his experimental results, they persuaded colleagues ready to believe that the force between hypothetical droplets of electric charge diminishes, as does gravity's, as the square of the distance between them. Coulomb did similar measurements on long bar magnets with widely separated poles supposedly containing magnetic fluids and got similar results: droplets of the same fluids repel, and of different ones attract, one another via an inverse-square force.

Coulomb's experiments date from 1785. By then other *physiciens-géomètres*—F.U.T. Aepinus (electricity and magnetism), Henry Cavendish (electricity, thermometry, pneumatics), P.S. de Laplace (heat, optics, capillarity), to name but one prominent representative from each of the leading academies (Berlin and St. Petersburg, London, Paris)—were quantifying the new domains of physics in analogy to gravitational theory. One of these domains—which, after 1770, rivaled electricity in public demonstrations—was pneumatics, which expanded following the discovery by Joseph Black, Joseph Priestley, and Cavendish that "air," which had been considered elemental since antiquity, consists of distinct permanent types. This capital discovery—which was analogous in form, as it was equal in importance, to Newton's detection of the composite nature of sunlight—created a productive overlap between physics and areas subsequently assigned to chemistry. The overlap widened around 1800 when the invention of the battery by Alessandro Volta (a Catholic professor in Pavia), and of a way to associate numbers with atoms by John Dalton (a Quaker school teacher in Manchester), demonstrated both the ecumenicity and accessibility of the then nascent physics.

Although the principles underlying the main branches of physics around 1800 were not reducible to one another, they bore analogies strong enough to encourage the belief that science might regain the sort of unity that Aristotelian concepts once gave *physica*. A "standard model" emerged that assumed the existence of two electric and two magnetic fluids, Newton's light particles, and caloric, a self-repellent fluid of heat that accounted for the elasticity of gases as well as for thermal phenomena. In addition, incorporating ideas developed by Jean-André Deluc (gadfly tutor to the Queen of England) and others who studied nature's grand open meteorological laboratory of the atmosphere, the model emphasized the interactions of caloric, light, and electricity among themselves and with the gas types. The special qualities of the active fluids, all conceived as weightless centers of distance force, displayed strong analogies, such as those between temperature and electrical tension, and between capacities for heat and electricity. The "imponderable standard model" thus had a plausible coherence, although through form rather than matter.

Despite being dominant around 1800, the imponderable model had competitors. One, explicitly opposed to the atomic picture, traced its roots to Immanuel Kant's *Metaphysische Anfangsgründe der Naturwissenschaft* (1786), which constituted matter and its properties from opposing forces of attraction and repulsion; physics, "in so far as there is mathematics in it," is the science of these forces. By almost pure thought F.W.J. Schelling developed a *Naturphilosophie* that extended the scheme of polar forces to all physical phenomena. Until 1830 Schelling and his followers held important chairs in German-speaking universities. *Naturphilosophen* less doctrinaire than he, like H.C. Ørsted (Copenhagen) and Thomas Seebeck (Berlin) sometimes entered the laboratory where, guided by notions of the unity and polarity of matter, they discovered connections between electric currents and both magnetism (Ørsted) and heat (Seebeck).

Institutional frameworks

The charge of the early modern college or university, which catered to students in their early teens, was disseminating rather than advancing knowledge. Teachers of physics ambitious for literary fame would write a textbook, not a research paper, and, if interested in showing experiments, would have to obtain their own instruments. Around 1750, however, universities and colleges began to acquire the apparatus of deceased professors at a discount and soon had the burden of maintaining and augmenting their bargains. The cabinet of instruments, with its director-professor, mechanic-servant, and small budget, thus became as much a part of a university as a library. Some research inevitably resulted, notably at Bologna and Pavia—showplaces, respectively, of the Catholic Church and Austrian Lombardy—and at Protestant Göttingen, the only significant university founded during the eighteenth century in Europe. Physics as a career began to take shape: in the early eighteenth century three out of four professors of physics left their chairs, usually to teach and practice medicine; in the late eighteenth century, the proportion was reversed, and most professors of physics retired from or died in harness.

Some of these persevering professors worked in newly established higher schools for engineering, mining, forestry, and the military, such as the royal artillery school in Mézières where Nollet taught and Coulomb studied. Fifty years later, the then new École polytechnique, a product of the educational philosophy of the French Revolution, taught the latest quantified physics and prepared the leaders and future professors of physical science in France. The tie between teaching in these technical schools and the systematization of physics may be symbolized by the four-volume *Traité de physique experimentelle et mathématique*, the first modern textbook of physics, published in 1816 by a graduate of the École polytechnique, Jean-Baptiste Biot.

The advancement of knowledge was the mission of academies. Their self-image may be inferred from their mottoes. The Royal Society's ran *Nullius in verba*, which the overly educated recognized as Horace's *nullius addictus iurare in verba magistri*, "bound to swear in no master's words." Thus they declared as their first order of business independence from their king, the teaching of the schools, and the dogmas of religion. The Paris Academy at its reorganization in 1699 took as its seal a sun, symbolic both of knowledge and of its king, Louis XIV, and for its motto, *Invenit et perfecit*. Its business was discovering and completing knowledge. The Royal Swedish Academy of Sciences, established in 1739, preferred to identify with an old man working the earth under the slogan, *För Efterkommande*, "for posterity."

Though dissimilar in structure and operation, the prototypical academies of Paris and London had in common the pursuit of natural knowledge exclusively. Most other academies also had sections for arts and *belles-lettres*. At least 60 such academies with substantial interest in natural knowledge were founded between 1700 and 1790. France led the way, with half the total, reflecting provincial competition with Paris and a felt need by professionals—lawyers, priests, physicians, and military officers—to express themselves outside the institutions of church and state. These academies helped to guide research in physics by judging papers for publication and awarding prizes and medals.

An analysis of the accomplishments for which the Royal Society awarded its prestigious Copley medal suggests the research topics esteemed by the academic system. Electricity and magnetism dominated Copley awards around 1750 just as they accounted for about a third of all the prizes in physics awarded by the French academies. Airs, heat, and geodesy made strong showings during the last third of the century. In all, the Royal Society chose achievements in experimental natural philosophy for two out of every three Copley medals it awarded through 1820, apparently regarding discoveries in physics as more fundamental than those in mathematics, natural history, anatomy, and medicine combined.

In contrast with astronomy—which, with the introduction of achromatic lenses around 1760 and the construction of large reflectors in the 1780s, produced discoveries of a quality not seen since Galileo (notably William Herschel's planet Uranus)—physics experiments typically did not take place in facilities specially built or equipped for the purpose. Most experimenters adapted rooms in their homes, and household implements, to their purposes. Sometimes they enlisted workshops, as William Lewis did for the study of platinum that earned him the Copley medal for 1754. In a cannon foundry in Bavaria, Count Rumford, Copley medalist in 1792 for his experiments on heat insulation, discovered in the grinding of gun barrels a powerful argument in favor of the kinetic theory of heat. Another quasi-industrial setting for experiments was the instrument-maker's shop.

Among reserved spaces for physics experiments, Martinus van Marum's deserves special mention. Practicing physician, public lecturer, curator of instruments, and academician, he had charge of a rich private institute for the cultivation of arts and sciences (the Teyler Foundation in Haarlem). Sharing the modern view of experiment enunciated by Priestley—"a person would stand the best chance of hitting upon some new discovery by applying a greater force than had hitherto been used"—van Marum commissioned the largest electrostatic generator made before the Van de Graaff machines of the 1930s. It did not repay the investment.

The great Teylerian electrical machine was the king of a line of lesser devices that, like it and the air pump, could serve both research and demonstration. Measuring devices, in contrast, were unlikely to entertain a crowd, and at first did not enlighten their users. They became more apt and exact after 1770, as appears from Volta's bottled straw electrometer, which was not only more precise than Nollet's earlier string electroscope, but also related more directly to the theoretical quantities it measured. Systematic work by Deluc, Cavendish, and others resulted in standardized accurate thermometers, barometers, and hygrometers. Routine meteorological and geophysical measurements, as of atmospheric temperature and pressure and earth tremors, inspired the invention of reliable recording instruments. The subtlest recorder employed in physical experiments in the eighteenth century was a frog. Using decapitated specimens Luigi Galvani (Bologna) detected the animal electricity that inspired Volta to make electrometers better calibrated than amphibians, to generate animal electricity without using animal parts, and to magnify the effect into an epoch-making prime mover: the Voltaic pile or battery.

Like every physical novelty during the eighteenth century, animal electricity or galvanism was enrolled in demonstrations.

Guillotined heads wired to a Voltaic cell winked gratefully in response. More impressively, the hanged murderer to whose spinal cord and muscles Dr. Andrew Ure of Glasgow attached galvanic leads showed rage, agony, and despair, and died again with a horrible smile.

Physics and enlightenment

Several factors operating in the wider society reinforced the acceleration of natural knowledge around 1750. Closest to the intellectual shift in physical sciences was an epistemology developed by the *philosophes* favoring the unlikely coupling of modesty and mathematics. A second factor promoting rationalization of natural philosophy was the increasing application of mathematics to the purposes of the enlightened governments of the later eighteenth century. The third factor was manufacture, as represented particularly by the trade in instruments. The *Encyclopédie* of Denis Diderot and Jean Le Rond d'Alembert, published for the first time between 1751 and 1772, made much of the epistemology of the *philosophes* and of the rationalization of the manufacturing arts. Toward the end of the century, rhetoric connecting physics and utility could emphasize the spectacular cases of the lightning rod and navigable balloons.

The French Enlightenment drew on natural knowledge for its central project of destroying organized religion. Voltaire's *Elémens de la philosophie de Neuton* (1738), the first comprehensive popular account of Newton's work on cosmology and optics in French, begins with Newton's notions of God and proceeds to correct thinking about free will, natural religion, and mankind's common humanity. Francesco Algarotti added enough of the philosophy of John Locke to his *Newtonianismo per le dame* (1737) to excite the Congregation of the Index. Pope Clement XII had then just condemned Locke's *Essay concerning human understanding* (1690), a master text for the Enlightenment, and the Holy Office had done the same for his *Reasonableness of Christianity* (in 1734 and 1737, respectively). The Catholic Church thus helped to recommend Locke's sensationalist psychology as a weapon against itself in the *philosophes'* campaign to crush it.

D'Alembert coupled Newton and Locke, the one the reformer of physics, the other that of metaphysics, in his lengthy "Discours préliminaire" to the *Encyclopédie* (1751). Together they overthrew Descartes' system, indeed all systems. Newton had demonstrated how far the human mind could soar when freighted only with mathematics and a few capital phenomena, and Locke had shown that all minds begin with even less, that is, with nothing at all. David Hume then worked hard at an experimental physics of the mind based on a power of association, an analogue of Newtonian attraction, which arranges sensations in accordance with their degree of resemblance and contiguity in space and time. No necessity links them.

Kant dug a deeper foundation for the proposition that the human mind cannot penetrate to the bottom of things. He made the concepts of space and time, which for Hume were deposits of experience, the conditions of perception. What the mind can acquire dependably is knowledge only of the appearances it arranges in space and time and orders, where appropriate, by cause and effect. Among the conclusions that Kant's mind derived from these principles was that physics must be quantitative: "in every special doctrine of nature only so much science can be found as there is mathematics in it." Hence, he wrote three years before the publication of Lavoisier's *Traité de chimie*, "Chemistry can become nothing more than a systematic art or experimental doctrine, but never science proper." A great philosopher need not be a great prophet.

The Enlightenment's program of rationalizing social institutions and artificial practices, and of sweeping away customs and concepts that had no deeper foundation than habit and tradition, coincided with the interests of bureaucratizing governments eager to curtail wasteful inefficiencies. The tougher, leaner, more mathematical experimental physics of the later eighteenth century both served and profited from this coincidence. Particularly after the Seven Years War, the larger states recognized the need for a better accounting, and more efficient use, of their resources. The entrance of the expert, the first step toward realizing the rationalization demanded by Enlightenment, the beginning of the effective mathematizing of natural philosophy, and the first glimmers of manufacture of items with interchangeable parts, were coeval.

Count Rumford—the American Tory who made his reputation in Bavaria designing nourishing soups and smokeless fireplaces—inspired the creation of an institution to transmit the latest science of soups, chimneys, and manure, the Royal Institution, which opened in London in 1799. It was to make central contributions to natural philosophy if not to scientific farming. Its second professor of natural philosophy was Thomas Young, reinventor of the wave theory of light. Humphry Davy, who became professor in his turn, remained faithful longer to the original program. His first important paper conveyed the results of his inquiries into tanning processes. Then, in a masterful exercise in grantsmanship, he obtained funds for a very large Voltaic battery on the ground that electricity enhanced the fertility of soil, and, with its aid, separated sodium, potassium, and calcium from their oxides.

"Science" in the eighteenth century often meant rational cut-and-try. It had its success in agriculture and among the inventive members of the informal Lunar Society of Birmingham, who met when the full moon illuminated their travels. Most of them also belonged to the Royal Society. The best known of their productions was the improvement of the steam engine by its member James Watt, who had learned something about heat and experimentation from Joseph Black when serving as curator of instruments at the University of Glasgow. The society thus supplied an early example of the academic–industrial complex as well as the machine that would propel the Industrial Revolution.

Because of the emphasis on mathematics in their curriculum, Jesuits sometimes occupied important posts for which it was requisite. For example, Ruđer Bošković (Roger Boscovich)—a mathematics professor in Rome and later in Pavia, who invented a *Theory of natural philosophy* (1758, 1763) requiring only a single universal distance force between point particles, and more usefully helped in repairing the dome of St. Peter's. He also made a trigonometric survey of the Papal States, founded the Jesuit observatory in Milan, and acted as director of optics for the French navy. The connection between the French navy and the Jesuits did not begin with him, however. In the last years of the seventeenth century, the French government established in several Jesuit colleges a royal chair of hydrography whose incumbent prepared students to find their way at sea.

The welter of German states and principalities, which abounded in universities and bureaucracies, and, after the Seven Years War, in problems of reconstruction, provided a theater for a special combination of economic rationalization and natural knowledge. *Cameralwissenschaft*, taught in universities and special higher schools, collected and integrated information about agriculture, commerce, industry, population, territory, history, laws, and customs to serve as a basis for economic policy. Perhaps its most successful and demanding subject was forestry, devoted to increasing sustainable yields by planting and harvesting trees in accordance with rationalized experience and the integral calculus.

The task of judging the health of the state by reading the "thermometer of public prosperity," that is, the census, brought natural philosophers close to the people, especially in France. Owing to the size and diversity of the population and its fear that a census would reveal its prosperity and raise its taxes, numbering the king's subjects was a difficult task. The confident new natural philosophers and mixed mathematicians did not think it beyond their reach. "Experiment, research, calculation are the probe of the sciences! What problems could not be so treated in administration! What sublime questions could not be submitted to the law of calculation!"

Fig. 20. A physics institute around 1900. The size, fit with local architecture, and arrangement of lecture halls, laboratories, and utilities at the University of Manchester were typical of institutes built around 1900. Rutherford was professor and director there from 1907 to 1919.

5

CLASSICAL PHYSICS
AND ITS CURE

During the nineteenth century, physics became weightier in several senses. The number of imponderables fell almost to zero. Physics became a recognized profession, whose British practitioners received (though they did not welcome) the name "physicist." In professionalizing, physicists acquired special training facilities in university institutes and technical schools that sprang up in Europe and America after 1870 primarily to train students as schoolteachers and electrical engineers (see Figure 20). During the last third of the century the perpetually shifting border between physics and mathematics moved again as physicists competent in contemporary mathematics set up as theorists; and again, as in the discovery of the gas types, concerns more chemical than physical brought in such fertile concepts as the ion and the table of the elements.

Nineteenth-century physics aspired to a theory of everything. A new standard model or project, mechanical reduction, replaced the program of imponderable fluids. The mechanical project penetrated more deeply than the imponderable one; for whereas the latter obtained superficial unity by describing many distinct substances in similar ways, the former premised an ultimate reduction to the same matter in diverse modes of motion. Once again Descartes did battle with Aristotle, accomplished a mighty synthesis, and prepared the way for something new. The physicists of the early twentieth century admired, and set aside, their nineteenth-century accomplishments as "classical," and built a quantum physics of atoms and molecules that, they claimed, contained all of chemistry, "in principle."

From the middle of the nineteenth century, physics increasingly made good on Bacon's promise that experimental science would improve the human condition. New industries competed to supply electric communication, light, and power, and to invent electrical appliances for the home and the workplace. New governmental agencies, national and international, came into existence to regulate services and standardize products. Old universities introduced new curricula to compete with new technical higher schools and replaced academies as the primary loci of research. At the same time, societies of physicists sprang up to see to the professional interests of their members.

Standard models

Drying up the fluids

The particles of light, with which the eighteenth century modeled Newton's rays, had had a rival since 1690, when Huygens explained double refraction by assuming light to be a pressure wave in a world-filling ether. No one developed a wave theory competitive with Newton's ray theory, however, until Thomas Young constructed one around 1800. He hoped to bring light and vision into parallel with sound and hearing, the subjects of the thesis that earned him a medical degree in Germany. Young's famous analysis of the interference patterns of light rays passing through two parallel narrow slits analogized the medium supposed to transmit light to a thin air capable of supporting longitudinal vibrations, and the patterns to the constructive and destructive combination of acoustic waves. The analysis met resistance in the phenomenon of polarization, which showed that light had a degree of freedom that a sound wave, characterized only by its phase, amplitude, and direction, could not represent.

The need to capture polarization in the model became urgent when French physicists, expecting to refute Young's theory, unexpectedly confirmed an apparently absurd consequence of it. The absurdity, discovered by Augustin Fresnel, a graduate of the École polytechnique, was that the patterns produced by light falling on a screen through a minute hole could have dark blots at their centers. The confirmation of Fresnel's deduction led to a concept more surprising than the effect it was invented to explain. Young and Fresnel independently suggested that the light-bearing ether could attain the needed extra degree of freedom if it oscillated "transversely," at right angles to the direction of the wave. To transmit transverse vibrations the "luminiferous ether" had to behave like an elastic solid rather than a thin air. Physicists fatigued themselves during the rest of the century imagining an ether that was rigid enough to propagate light yet soft enough to offer no resistance to planets. Despite the problems of modeling space as a solid, most knowledgeable people had dropped the light particle in favor of ether vibrations by 1830.

At the same time, and following a similar pattern, the magnetic fluids also disappeared. Again an experimental demonstration made on the periphery (Paris then being the center of physics) prompted a French mathematician to a destructive generalization. The actors were H.C. Ørsted (Copenhagen), who discovered that a current-carrying wire could act at a distance on a magnetic needle (1819), and A.-M. Ampère (Paris), who showed that a circular current behaves as a magnet and deduced that magnetism arises from electricity in motion (1822, 1827). This model implied the existence of distance forces as awkward as an elastic-solid vacuum: the magnetic force of a current lies not in its direction, but in circles around it, and the force between moving droplets of electrical fluid depends on their relative velocities as well as on their separation. Furthermore, to include the generation of an electric current by a changing magnetic force (1831), the most famous of the many famous discoveries of Michael Faraday (Royal Institution, London), the force-law must also involve the relative acceleration of the droplets. In an ironic reversal of polarity, the proponents of these peculiar distance forces were continental physicists, notably Wilhelm Weber (Göttingen) and Hermann von Helmholtz (Königsberg, later Bonn and Berlin), whose predecessors had regarded Newton's gravitational acceleration as unintelligible, whereas Newton's countrymen, Faraday and his followers, tried to do without distance forces altogether.

Faraday eliminated the force between electrical droplets by discarding the droplets. Instead, he located electricity in a special medium he supposed to exist in the space between electrified bodies. The stresses and strains in this medium or "field" constituted, and conveyed, what appeared to ordinary physicists as electric and magnetic forces acting at a distance between "charged" or "magnetized" bodies. William Thomson, later Lord Kelvin (Glasgow), translated Faraday's intuitions into a dynamical picture portraying magnetic forces as vortices, and electric forces as linear flows, in the field. In the 1860s James Clerk Maxwell (Aberdeen and London, later Cambridge) worked out a full dynamical model of Faraday's field-medium capable of representing most known electrical and magnetic phenomena. Using the model as an expendable scaffold, he raised the edifice of an enduring electrodynamics, a set of partial differential equations ("Maxwell's equations," 1866) linking electric and magnetic forces and their sources. Thus, for those who followed the Faraday–Maxwell line, electrical fluids had no better claim to existence than magnetic fluids or corpuscles of light.

Nothing, therefore, could have been more gratifying to the grand-unifying physicist than to learn that Faraday's field and the luminiferous ether were one and the same. In 1864, Maxwell calculated from his model and known electric and magnetic parameters that the speed with which the hypothetical electromagnetic disturbances travel through space came close to the latest measurement of the speed of light. In 1887 Helmholtz's former student Heinrich Hertz (Karlsruhe) generated and detected the electromagnetic waves Maxwell's theory predicted. The resultant grand synthesis complicated the task of ether-field theorists, who now required an elastic-solid medium filling all space, competent to produce all the phenomena of electricity, magnetism, and light, and transparent to gravitating bodies.

Like light particles, caloric had had a challenger in the eighteenth century that supposed heat to be a mode of motion. Count Rumford's notorious production of heat from grinding cannon muzzles, though pertinent to the challenge, had produced more noise than light. He was prescient, however, in locating his experiments in an industrial setting. The decisive experiments against the caloric theory, which date from the 1840s, took place in a family brewery whose owner, James Prescott Joule, once a student of John Dalton, used exact thermometry for product control. Applying his instruments to engines, he showed that they consumed heat in raising a weight and that a certain quantity of mechanical effort, or a certain expenditure of zinc in a battery, always generated the same amount of heat. Whatever heat was, it could be created and destroyed. This conclusion, based on delicate measurements, had to overcome not only skepticism about Joule's thermometers but also a brilliant theory of steam engines.

This theory, which applied to all engines, regardless of fuel, mechanism, and working substance, assumed the indestructibility of caloric. Its author, Sadi Carnot, a *polytechnicien* like Fresnel, calculated the maximum efficiency possible for any heat engine as a function of the temperatures of the boiler and condenser. He argued that to escape a possible perpetual motion, the most efficient machine must be reversible; and to be reversible, its operations must never allow contact between parts maintained or working at different temperatures. From the ideal cycle of operations he invented to satisfy this condition, he deduced that the engine's efficiency, the ratio of the weight lifted to the amount of caloric fluid Q passing through the machine, should be proportional to the difference of temperature ΔT between the boiler and the condenser.

The clarity, ingeniousness, and plausibility of Carnot's analysis of 1824 impressed William Thomson. But he also inclined to Joule's theory, which, in contrast to Carnot's, destroyed caloric in doing mechanical work. Mathematics indicated a path to reconciliation. Thomson saw that Joule's expression for efficiency would be the same as Carnot's if the conserved quantity was not Q, but Q/T. Two principles, not one, were involved: (1) mechanical work and electricity could be converted into heat and vice versa without loss (Joule) and (2) a heat engine working reversibly conserves the quantity Q/T (Carnot). Spying difference in sameness (to speak with Timaeus) can be as productive and difficult as identifying sameness in difference.

Rudolf Clausius (Bonn) also reconciled Carnot's conservation theory with Joule's transformation theory, which Clausius had in the form developed independently of Joule by the physician Robert Julius Mayer. Clausius called the coin of conversion among thermal, mechanical, electrical, and other forces of nature "energy." The mysterious quantity Q/T—nature's fee for the conversion—he called "entropy." In the ideal reversible case, the fee is zero. In practice, small amounts of heat are lost so as to render all "real" processes irreversible. In Clausius' formulation, the energy of the universe stays constant while its entropy strives for a maximum ("the laws of thermodynamics," 1865). Thus considerations arising in industrial settings drove the most plausible and ancient of the imponderable fluids, the matter of fire and heat, out of physics. It is but a mode of existence of molecules.

Back to matter and motion

The concept of heat as motion proved as far-reaching as the synthesis of light and electrodynamics. It made possible a quantitative link between mechanical concepts and temperature via parameters that measured molecules and allowed mathematicians passage between the macro and microworlds. The wormhole to this wonderland lay in the equation for a "perfect gas," a legendary medium described by Boyle's law as generalized by Laplace's protégé J.L. Gay-Lussac (Paris): pressure = $(R \times \text{temperature})/\text{volume}$. The constant R was the wormhole. A simple model ("the kinetic theory of gases"), in which the pressure exerted by N molecules of perfect gas arises from their bombardment of the walls of the vessel containing them, gave an astonishingly simple "law": the *vis viva* (or kinetic energy) of a perfect gas molecule is $3kT/2$, where k, the bond between mechanics and thermodynamics, and the measurable and the molecular, is R/N.

The simple theory that related *vis viva* to kT ignored collisions among gas molecules, distributions in velocity, mean free paths, and, most importantly, a statistical treatment of equilibrium. Clausius, Maxwell, and Ludwig Boltzmann (Graz and Vienna) added these realistic touches, and, after heroic calculations, recovered the old result: each degree of freedom of motion of each molecule has an energy $kT/2$ at equilibrium ("the equipartition of energy"). Molecules able to move only in translation in three dimensions have an average energy of $3kT/2$. The specific heat of a mole of gas whose N molecules each has f degrees of freedom would be $(f/2)R$. The formula worked well—for monatomic molecules.

Heat equilibrium occurs when entropy is a maximum. This simple statement hides a deep difficulty. If a gas is a perfect mechanical system, it should be able to move out of equilibrium as easily as into it, since the laws of motion run equally well backward as forward. Hence a statistical-mechanical representation of entropy appeared impossible to many physicists. Boltzmann argued that, owing to the colossal number of molecules in play, the equilibrium distribution is overwhelmingly the most probable one; departures from it will almost certainly be reversed instantaneously (1872, 1877, the "H-theorem"). Entropy consequently could be understood as the probability—or, to be exact, as proportional to the logarithm of the probability—of finding the system in a particular state. The proportionality factor k, now appropriately called "Boltzmann's constant," links not only the macro and the microworlds, but also the living and the dead. It appears on Boltzmann's tombstone in the form $S = k \log W$.

The successes of the gas theory and Maxwell's electrodynamics, the equivalence of all energy to mechanical energy underwritten by the first law of thermodynamics, and the relative ease with which physicists reasoned with the intuitive concepts of matter and motion combined to return nineteenth-century physicists to the dream of Descartes. The president of the French Physical Society, Alfred Cornu (Paris), opened the first international congress of physics, held in Paris in 1900, with the reassurance that Descartes "hovered" over the meeting. The great mathematician Henri Poincaré (Paris) gave the keynote address. He advised his audience that they should have no special fondness for any world system. Rather, their task was to collect the facts of experiment and arrange them for consultation in the most convenient manner. In this, he said, physics resembled librarianship more than philosophy.

Poincaré's unflattering estimate of the epistemological status of physics corresponded to the considered opinion of many physicists. Although they might act and speak as if they sought the truths of nature, when they philosophized they acknowledged that the goal was beyond them. A prime impulse to this neo-skepticism was the demotion, by Gustav Kirchhoff (Heidelberg), of the most secure branch of physics, analytical mechanics, from a true account of matter and motion to a mere description of it. This "descriptionism," to use a Victorian phrase, turns up in the influential epistemologies of physicists who took a broad view of their discipline, notably Ernst Mach (Vienna), and also Joseph Larmor (Cambridge), who sponsored an English translation of Poincaré's writings.

Physicists as librarians

The metaphor of the library suited a large part of nineteenth-century physics, which boasted many new laws or effects easily entered into Poincaré's imaginary catalogue. Representative entries concerning metals might read, alphabetically: "Electric current, generated by heating a junction of dissimilar metals," Thomas Seebeck, 1822; "Heat capacity, atomic, inversely proportional to atomic weight," Pierre Louis Dulong and Alexis Thérèse Petit, 1819; "Heat conductivity, proportional to T times electrical conductivity," Gustav Wiedemann and Rudolf Franz, 1853; and "Paramagnetism, proportional to $1/T$," Pierre Curie, 1895. The inverse of these effects also had entries, for example, "Heat developed by an electric current passing a bimetallic junction," J.C.A. Peltier, 1834.

The library added new shelves with the discoveries that mechanical forces could create electricity (piezoelectricity) and magnetism (magnetostriction). In 1888, in his *Applications of dynamics to physics and chemistry*, J.J. Thomson (Cambridge) showed that vast new additions were possible, at least theoretically. For example, if twisting a wire while warming it generates an electromotive force, then putting a current through a warming wire should twist it. Pursuing such reciprocal effects, James Alfred Ewing, who established the frontier between physics and engineering at Cambridge, discovered hysteresis, the lagging of magnetic effects behind their immediate causes (1882–5).

The most exact and detailed entries in the library of physics around 1900 were the characteristic wavelengths of the series of spectral lines emitted by the elements when heated sufficiently. The first such survey, made by Kirchhoff and Robert Bunsen (Heidelberg) in 1860, employed salts vaporized in Bunsen's burner. The value of their catalogue appeared immediately in Bunsen's discovery of cesium and rubidium, so named after the strongest colors in their spectra. Many scientists then joined in measuring and classifying the lines of the elements, which opened a way to explore the constitution of stars as well as of substances here below. Librarians of spectra invented several cataloguing rules, notably the "Balmer formula" (after Johann Balmer, 1885), which reduced the wavelengths of a series of hydrogen lines to an empirical constant and a running integer, and its generalization, by Johannes Rydberg (Lund, 1888), to many series and elements. Their formulations contained the clue to what the Greeks would have rated an oxymoron: the internal structure of atoms.

Meteorologica provided many important new entries for the physics catalogue of 1900. In confirmation of Deluc's brave assertion that meteorology is the most basic of the physical sciences, Dalton had endowed atoms with different sizes to explain why the atmosphere's constituents do not separate out according to weight. The different sizes, which implied different weights, cause each gas type to act as if the others were not present. After much wrangling over atomic weights, chemists came to agree enough in the 1860s to prompt the creation of a "periodic table," a Ouija board that ordered known elements into families by weight and predicted the existence of unknown elements soon discovered. In the 1890s meteorology again intervened significantly in atomistics when Lord Rayleigh (formerly Cambridge) discovered a minor constituent of the atmosphere he named argon. A family of "noble gases" of the inactive argon type beginning with helium soon disclosed itself to the chemist William Ramsay. The new family fitted into the Ouija board perfectly, with one exception: argon and potassium had to swap places rightfully theirs by weight to preserve chemical periodicity. The noble gases thus confirmed the power, and deepened the mystery, of the periodic table of elements.

By 1800 progress in hydraulics and thermodynamics made possible a vision of a dynamical analysis of weather, whose realization, however, would not come for 150 years. In contrast, Newton's tidal theory, refined by treating each component of the generating forces separately, opened itself to effective though tedious "harmonic analysis" carried out under William Whewell (Cambridge) and much improved by Lord Kelvin. Tidal theory grew powerful enough to become seriously misleading. The prevailing theory of the Earth's formation around 1850, the "nebular hypothesis" started by Kant and Laplace, derived the Sun and its planets from a spiraling gas cloud that heated as it coagulated. While the Earth cooled from its original molten state, its exterior formed a crust that crumpled as the molten rocks under it cooled and contracted. To account for the migration of the Earth's magnetic poles (the common explanation of the variation of the compass needle), physicists supposed that the crust moved relative to a liquid core. Kelvin observed that a mass of liquid under the crust should show tidal effects. None being known, he concluded that the Earth is a rigid body.

From thermodynamics and the properties of surface rocks, Kelvin calculated that the Earth has been cooling for more than 100 million years—an indigestibly long time for people who credited biblical chronology, but not long enough to give Darwin's evolution time to work. Perhaps more appealing to biblical literalists, Kelvin did not allow the human race very long to improve itself. The second law, he asserted, insured that the Earth, once too hot for humans, soon would become too cold for them. His widely accepted views about the age and rigidity of the Earth, apparently anchored on rock-hard physical principles, did not survive long into the twentieth century. They failed in two steps. Analysis of seismic waves from deep earthquakes showed that only longitudinal waves crossed the core and that they traveled more slowly there than elsewhere. This information allowed Beno Gutenberg (Göttingen, 1912) to estimate that the radius of the core is almost half the Earth's radius, and everyone to infer that, since liquids do not support transverse waves, the Earth is not solid throughout. By similar techniques, Andrija Mohorovičić (Zagreb) located a discontinuity some 30 miles beneath the continents that divides the crust from the "mantle" (1909–10). Thus macroscopic physics softened Kelvin's rigid Earth. As for his age of the Earth, microphysics, in the form of atomic energy, exploded it beyond the needs of evolutionists.

Woes and wonders in 1900

Kelvin's clouds

Kelvin was an exemplary physicist not only because of the importance of his discoveries, his breadth of interest, and his dedication to the program of mechanical reduction, but also because he could diagnose serious difficulties in his science. In 1900 he bundled his reservations into two problem areas, which, in happy reference to meteorology, he labeled "clouds." One he seeded on the fact that no one had managed to design an ether that could do everything Maxwell required of it. The other was the equipartition of energy.

Like Kelvin, Boltzmann acknowledged that the progress of mechanical reduction had its difficulties, but saw no other way forward. He ridiculed the physical chemist Wilhelm Ostwald (Leipzig), the champion of a science based only on energy and its transformations ("energetics"), for admonishing colleagues, as Moses had the Hebrews, not to "make unto thee any graven image, or any likeness of anything." This commandment (Exodus 20:4–6) succeeded no better among physicists than it had among the worshippers of the golden calf. Most continued to flirt with molecules. Others, however, like Max Planck (Berlin), although critical of energetics, stuck as close as they could to the two laws of thermodynamics. Still others thought to reverse the argument that gave priority to mechanics and substitute electricity or heat in its stead. Boltzmann associated these schisms with the *fin de siècle* attack on received canons of art, music, and literature. Everywhere, he said in 1899, classicism had its enemies. But for clarity, longevity, and productivity he would stick to what he, perhaps the first among mortals, called "classical physics"—the physics of mechanical reduction, the science of clear pictures in space and time—and trust that some imperial Newton would dispose of equipartition.

Kelvin's clouds grew more thunderous after 1900. Through Maxwell's grand synthesis, the problem of how light propagating in the omnipresent stationary ether affected a body moving through it had taken on new urgency. Theory explained some optical phenomena by supposing that moving matter drags a portion of the ether with it; but when the ether became electromagnetic, and the same theory required that it be pulled by spinning charged disks, it stubbornly remained quiescent. And yet moving ponderable bodies seemed to carry the ether surrounding them, since experiments to measure their velocity through it invariably failed. The most famous and delicate of these experiments, by A.A. Michelson (Cleveland, later Chicago) and Edward Morley (Cleveland), date from the 1880s.

As when Galileo sliced through the accumulated conundrums of motion by replacing physics with mathematics, so now H.A. Lorentz (Leyden) transformed Maxwell's equations for moving bodies so as to kill terms that predicted detectable effects arising from the motion. His maneuver in effect did away with the ether as a mechanical entity. In 1905, Albert Einstein, then a patent examiner in Bern, recognized that the Lorentz transformations, which Poincaré had also derived, were just what he needed to describe the physics in bodies moving inertially—that is, with constant rectilinear velocities—relative to one another. He did without ether and shelved the exotic problem, then agitating many theorists, of whether electrons changed shape when moving.

Special relativity (Einstein's theory of 1905) is as democratic as equipartition since it effectively places every observer at rest within his own ether. Light consequently always propagates toward or away from the observer in free space at the same speed irrespective of his state of motion. To this principle Einstein added the equally intuitive proposition that the laws of physics should be the same in all inertial frames; no physics experiment can tell you whether the passing train or yours or both are in steady motion. From these easily comprehensible beginnings Einstein deduced very bizarre consequences: time dilation and space contraction (clocks run slower, and sticks grow shorter, in "moving systems"), and the equivalence of matter and energy. However disorienting the conclusions, the principles could be construed as a rational expansion of the principles of classical physics.

Discharging Kelvin's second cloud required more radical thought—even a passage through irrationality. The problem that provoked the madness concerned the equilibrium distribution of radiant energy (as a function of frequency or color) contained in an oven at constant temperature ("blackbody radiation"). This seemingly obscure problem had some industrial interest, as it related to standards of illumination. There was no reason to believe it would be intractable, and the library of physics already had a place for its solution on the shelf between electrodynamics and thermodynamics. Equipartition obtruded, however, whenever the ether was treated as a mechanical system; and equipartition implied that every mode of ethereal vibration received the same amount of energy, $kT/2$. As emphasized in 1900 by Lord Rayleigh, this democratic division could be catastrophic. If, like other vibrating systems, the ether had many more modes of vibration at high frequencies than at low, all the electromagnetic energy available to it should run into the ultraviolet and beyond.

To escape this predicament, Rayleigh and, more assertively, James Jeans (Cambridge) suggested that the process might take a long time—longer, perhaps, than the age of the universe—and so saved equipartition at the cost of postponing equilibrium to a time when no one was likely to observe it. That was not reasonable. Knowing nothing of Rayleigh's deliberations, Planck published in the same year, 1900, a theoretical formula for the blackbody spectrum that appeared to agree with experiment. Soon, however, measurements in the infrared negated the formula. Planck thereupon found a new one, which worked, and a new theory for it, which subverted the classical physics of Boltzmann on which he thought he had based it. That was irrational. Planck's "quantum theory" soon joined with other evidence that the microworld could not be described using the ordinary concepts of physics.

The microworld

William Whewell, who gave the world "physicist" and "scientist," supplied Faraday with "ion" to refer to the electrified unknowns in motion in a working electrolytic cell. In his doctoral dissertation at the University of Uppsala in 1884, Svante Arrhenius identified ions with charged molecular fragments and claimed their presence in all solutions. His concept of dissociation, with its counterintuitive consequence that a strong solution of a good electrolyte contained few solute molecules, proved its power in fields far from its origin, particularly in the study of electrical discharges through dilute gases. During the 1870s and 1880s, physicists discovered "cathode rays," which proceed in straight lines from the cathode of the discharge tube to disclose themselves by causing the opposite walls to fluoresce. (Like "ion," Whewell made "cathode" and "anode" for Faraday, who wanted descriptive terms uncontaminated by electrical theories.) The obvious hypothesis, which supposed the rays to be negative gas ions repelled from the cathode, failed in one essential point. J.J. Thomson and others showed around 1897 that cathode-ray particles must have a ratio of charge to mass (e/m) about 1,000 times as large as that of the lightest electrolytic ion, hydrogen. Thomson interpreted this result to mean that the

Fig. 21. Desktop physics. W.C. Röntgen pictured in his laboratory in Würzburg in 1895, the year he discovered X-rays. Research spaces in the institutes were few and small around 1900, although many physics experiments required dependable electricity and good vacuum pumps.

cathode-ray "ion"—or, in his language, "corpuscle"—represented matter in a state of complete dissociation. Implicitly his analysis marked an end to the Faraday–Maxwell program of doing away with a distinct material of electricity.

The recognition of the corpuscle/electron capped a series of extraordinary discoveries associated with the gas-discharge tube. Late in 1895 Wilhelm Conrad Röntgen (Würzburg, later Munich) inferred from the glow of a phosphorescent screen some distance from a discharge tube generating cathode rays the existence of penetrating radiation of novel character (see Figure 21).

He could not reflect or refract its rays or bend them in a magnetic field; in compensation, he could use them to photograph the inside of a living human hand. He summed up his knowledge of their nature by calling them "X-rays." Following up an analogy to fluorescence, Henri Becquerel (Paris) found that uranium gave off rays different from Röntgen's. Marie Curie and her husband Pierre Curie (Paris) discovered in 1897–8 that thorium and the elements polonium and radium, which they discovered in uranium ores, also had the ability to radiate spontaneously ("radioactivity").

Thomson and his student Ernest Rutherford established that all the new rays ionized the air, which enabled Rutherford to distinguish a soft ("alpha") and a more penetrating ("beta") component among the Becquerel rays. The e/m of the beta ray came out close to that of the cathode-ray particle. Thomson's inference that electrons might be the building block of matter had further confirmation in the experiments of Pieter Zeeman (Leyden, then Amsterdam) as elucidated by Lorentz. In 1896 Zeeman managed to split some spectral lines in a magnetic field (the "Zeeman effect"). Lorentz's calculation of the magnitude of the splitting required that the electric oscillator supposedly responsible for the lines have an e/m close to that of the corpuscle. Thus, around 1900, physics got its first elementary particle and half a dozen rays for which its library had no shelf mark.

Continuing his study of alpha particles, Rutherford, now a professor in Montreal (1902), and his chemist colleague Frederick Soddy uncovered the tendency to self-destruction of radioactive substances. Most of the many fleeting decay products subsequently detected did not fit into the periodic table. With this information, radiochemists and atom builders realized, by 1913, that the discrepancy at argon/potassium was systematic: the Ouija board worked correctly when ordered not by weight but by an integer beginning with unity at hydrogen and assigned on the basis of chemical properties. The realization allowed more than one radioelement to occupy the same cell in the table ("isotopy"): atoms differing in weight could have identical chemical properties. As a discovery, isotopy ranked in logic if not in depth with thermodynamics: in each case scientists realized that a single concept (atomic weight, energy) required another (atomic number, entropy) to provide an adequate description of the phenomena.

Shortly before the recognition of isotopy, Rutherford invented his version of the nuclear atom. It deviated from Thomson's market-leading model, in which electrons circulate within a positively charged space, in being unstable both mechanically and electrodynamically. That had ruled it out in its previous versions. But by 1911 the few bold spirits who had given up on ordinary physics in the microworld could think that Rutherford's atom was so bad that it might be a good place to look for quantum activity. Its advantages included, besides having classical physics against it, offering a simple representation of Z, the atomic number, as the charge on the nucleus, and explaining some details about the passage of alpha particles through matter.

We may date the definitive entrance of physicists into the microworld to the last few years before World War I, when their increasing ability to assign exact values to the charges, masses, and numbers of atoms and electrons established their belief in it. The charge on the electron was the most valuable value: once in hand it produced, via measurable quantities, the electronic mass m, N, k, and, via k, Planck's quantum h. In 1910 Robert Millikan (Chicago) gave the value $e = 4.891 \times 10^{-10}$ esu (electrostatic units) in his "oil-drop experiment," a refinement of a technique using water invented in 1898 by J.J. Thomson. Planck's theory of blackbody radiation gave e as 4.69×10^{-10} esu, close to Rutherford's value from counting alpha particles, 4.65×10^{-10} esu.

Jean Perrin (Paris) confirmed these values from another direction by obtaining N from measurements of the dance of uniform tiny gumballs suspended in water between the tugs of gravity and osmotic pressure. Their dance, Brownian motion, arises from imbalances in the impacts of water molecules on them. In 1905 Einstein had calculated the mean displacement of the gumballs over time from the places at which they are first seen. The formula included N and quantities Perrin could measure or determine. His average, $N = 7.0 \times 10^{23}$, which he gave in 1909, agreed fairly well with 6.2×10^{23}, the value calculated using e determined by Rutherford.

Planck's successful formula for blackbody radiation contained two constants. One the theory identified as k; the other was an ad hoc number, h, which placed a threshold on the energy ϵ that an ether mode of frequency ν could take up. Hence "$\epsilon = h\nu$," a slogan as familiar to physicists as "$E = mc^2$." From spectroscopic measurements Planck could deduce good values for k and thereby for e and N. That his quantum hypothesis hid a breach with classical theory did not come fully to light until Einstein and others exposed it around 1905. Two years later, Einstein brought Planck's formulation closer to the classical problem of equipartition by applying it to the vibrations (the specific heats) of elastic solids. According to the empirical law of Dulong and Petit, metals all have the same specific heat per mole independent of temperature. Applying Planck's radiation law, Einstein obtained specific heats that deviated from this constant value at low temperatures. Measurements by Walther Nernst (Berlin) confirmed Einstein's extrapolation and the insight that Planck's law applied to vibrations far removed from its original jurisdiction. The problem of the quantum moved up on the agenda of physics. At Nernst's suggestion, the Belgian industrialist Ernest Solvay, who had his own scheme for replacing classical physics, sponsored a meeting in Brussels in 1911 at which all the important physicists interested in radiation and quanta came together to try to explain to one another what Planck's quantum h meant. They did not succeed.

Instances of irrationality multiplied. X-rays seemed to act like light sometimes and like non-light at other times. They spread like waves but interacted with matter like particles. It was as if (the image comes from W.H. Bragg, Leeds, 1910) ripples caused by a log dropped into water encountered an identical log floating some distance away, imparted all their distributed energy to it, and propelled it to the height from which the original log fell. The same paradox appeared in the photoelectric effect, discovered by Hertz in 1887, in which ultraviolet light, which everyone knew to be a wave, knocks electrons out of metals as if (so Einstein pointed out in 1905) it was composed of particles. Radioactivity offered another instance of paradox. Thomson had made the reasonable suggestion that a gradual loss of energy by atomic electrons through radiation caused radioactivity. Disciples of Boltzmann pointed out that Rutherford and Soddy's law of radioactive decay admitted the interpretation that in a small interval of time every atom of a radioelement has the same probability of exploding. Franz Exner (Vienna) drew the surprising conclusion: the probability of radioactive disintegration and other fluctuations were chance events incalculable in principle. Many physicists rejected the implied limitation on causality. Planck spoke for them all when, in the fateful August of 1914, he inveighed against those who took the easy way and denied the existence of strict causal laws just because some physical phenomena appeared to elude them.

A postdoctoral student working in Rutherford's laboratory in Manchester, Niels Bohr, welcomed the impasse. His doctoral thesis had generalized the electron theory of metals, pioneered by Thomson, Lorentz, and Paul Drude (Berlin), who treated conduction electrons confined in a wire as a free gas apart from their collisions with metal molecules. Bohr came closer than they did to deriving the Wiedemann–Franz law but failed with Curie's law and heat radiation. He placed the blame on equipartition and supposed that a restriction like Planck's must be imposed on atomic electrons. That did not surprise him. His reading of Danish philosophy and literature had prepared him to expect that physical theory must occasionally hit immovable barriers. For him, the conundrums of radioactivity and quanta were opportunities, not disasters.

Bohr seized his opportunity by accepting the nuclear atom and removing its defects by fiat: in their ground states, atomic electrons shall shirk their classical obligations to radiate and to perturb their common circular motion. Early in 1913 Bohr encountered the Balmer–Rydberg formula in the form $v_n = K(1/2^2 - 1/n^2)$. Multiplying both sides by h and reading the result as an energy equation in Planck's style, Bohr recognized the existence of excited "stationary states" in which electrons circulated with the immunities they had in the ground state and with energy $-hK/n^2$. Balmer lines originate irrationally in an unmotivated transition ("quantum jump") from the nth to the 2nd stationary state. By interpreting $-hK/n^2$ as the energy of the nth stationary state and allowing ordinary physics inconsistently to determine the relationship between the electron's velocity and nuclear distance while in the state, Bohr derived Rydberg's K in terms of the atomic constants $(K = 2\pi^2 m e^4/h^3)$ and sorted out attributions of spectral lines to hydrogen and helium. This tour de force persuaded Einstein, Jeans, and other alumni of the Solvay Council that Bohr had found a way forward.

Although Bohr offered a "correspondence principle" that hinted at how the calculations of his quantized atom could be made to agree with those of classical physics in certain limits, he emphasized that even where calculations might agree, the processes did not. His quantum jumps thus joined radioactive explosions as the first examples of natural phenomena placed formally beyond the reach of human inquiry since the time of Thales.

The profession

Most of the 700 attendees of the International Congress of Physics of 1900—about one in four of the world's physicists—

held paid posts in teaching (60 percent), industry (20 percent), and government (20 percent). An important novelty was the slowly growing subset of theorists. They occupied chairs primarily in Germany and Austria. In Britain and the British Empire, graduates of the Cambridge regime in mathematics held about half the permanent positions in physics in 1900. Though not theorists in the German sense, since few of them tried to make grand, quantified world systems, their ability to devise and calculate the behavior of mechanical models gave English physics an orientation that favored the production of atomic models like those of Thomson, Rutherford, and Bohr.

French physics had a quite different but equally distinctive epistemology, imbued with positivism and instilled at the École polytechnique and École normale, through which over half the academic physicists active in France in 1900 passed. This positivism, which expressed itself in reluctance to commit to models of the microworld, marked the approach of the Curies. No standard epistemology or hegemonic center existed in Germany, where students customarily attended more than one university or polytechnic, or in the United States, whose professors came from a variety of universities and typically finished their training in decentralized Germany.

Most informed observers ranked Germany first among the science-producing nations around 1900 largely because of the quality of German-language publications, the preponderance of German scientific instruments, and the generous support of science institutes by several states of the Reich. Decentralization and competition were the driving principles and the guarantee that in Germany the best scientists would not be crushed, as in

France, into the apex of a single educational system. Once at the Parisian apex, however, a *physicien* could regain space by the practice of *cumul*, or the accumulation of posts. Henri Becquerel, by no means an overachiever, held three professorships in Paris, one won on his own and two passed down as if private property from his father and grandfather.

In the British Empire little support for academic physical science came from the central government, and in the United States none. A large source of funds for expansion in both came from individuals or corporations. Spokesmen in each country, particularly in Britain, compared this uncertain funding unfavorably with the generosity of the German *Länder*. This perception missed the fact that with private philanthropy the material base for physics then was expanding more rapidly in the Anglo-Saxon countries than in their principal rival. German statesmen of science pointed to this generosity as something to fear and emulate.

During the decade before World War I, commercial, military, and rhetorical competition became ever more important for recruiting manpower and building institutes. Britain and the United States, embarrassed to have to send their products to Germany's Physikalisch-Technische Reichsanstalt (PTR, 1887) for certification, established standards laboratories of their own. Like the PTR, the British National Physical Laboratory (1900) and the American National Bureau of Standards (1901) employed graduate physicists in increasing numbers. Naturally they pressed to do research, some of which, like measurements of the blackbody spectrum at the PTR, related to fundamental problems. A few industries producing the products for testing also established research laboratories. Staff at the General

Electric Research Laboratory (1900) numbered perhaps 200 in 1913, including the former academic physicists Irving Langmuir and W.D. Coolidge. A little behind General Electric came American Telephone and Telegraph (1907), Corning Glass (1908), Eastman Kodak (1912), Philips Eindhoven (1914), and Siemens & Halske (1900, 1913). Most of them served the "goddess of the age," electricity (see Figure 22).

Three other institutional forms created with government and/ or industrial money in the early twentieth century affected the

Weltausstellung zu Paris: Die Wasserkünste vor dem Elektrizitätspalast auf dem Marsfelde.
Nach einem Aquarell von Ewald Thiel.

Fig. 22. The palace of electricity. The palace, a feature of the Paris Exposition of 1900, impressed visitors with its displays of colored lights. Electricity also won surprised friends by powering moving sidewalks around the extensive exhibition grounds.

tone and pace of physical science. The Nobel Prizes, endowed with the proceeds of dynamite and smokeless powder, were intended by their founder to reward those "who, during the preceding year, shall have conferred the greatest benefit on mankind." Although the first prize in physics, to Röntgen for X-rays, met this test, the professors soon conquered the system and rewards went more and more to academic work of no immediate practical value.

Andrew Carnegie's endowment of $10 million in 1901 for a research institute in Washington, DC, staggered contemporaries: interest on the money for one year probably exceeded the total expenditure on the laboratory of the Royal Institution for a century. Most of the income supported studies of meteorological subjects. German planners, concerned to inspire similar generosity among the Reich's industrialists, proposed the creation of a Kaiser-Wilhelm-Gesellschaft to assist in financing "pure" research institutes. Starting from a premise that reads oddly today—"Science has reached a point in its scope and thrust that the state alone can no longer care for its needs"—the society's projectors pointed to the Carnegie Institution, the Nobel Institution, and a gift of a million dollars to the University of Chicago for a physics institute, as proof that Germany was falling behind. The Kaiser-Wilhelm-Gesellschaft incorporated in January 1911, with a pledged capital of around a fourth of the endowment of the Carnegie Institution. The society's institute for physics opened virtually, with Einstein as director and Planck as effective executive, in 1917, but did not acquire a building for another 20 years, and then the Rockefeller Foundation paid for it. Apparently, the society did not consider an establishment run by Einstein and Planck likely to contribute much to German industry.

As physics professionalized, the old division of labor between the academy and the university ceased. Research increasingly became an expectation, and then an obligation, of professors. Careers depended more and more on publication in the transactions of the main scientific academies and professional societies. Among the societies, themselves a creation of the later nineteenth century, those of Berlin and London, and the Institute of Electrical Engineers, were perhaps the most prestigious publishers. Among the national academies, the French, British, and, far behind in status, the American perforce outdid the Germans, since Germany had no national academy. Regional ones remained important, however, especially those of Berlin (Prussia), Göttingen (Hanover), Leipzig (Saxony), and Munich (Bavaria). In addition, general or disciplinary news journals like *Nature* (founded 1871), *La nature* (1873), and *Physikalische Zeitschrift* (1899) disseminated research reports and good tidings.

Above the local level (universities, polytechnics, government and industrial laboratories) and the regional/national level (professional societies, academies) stood, in the scale of inclusiveness, the national associations for the advancement of science and international organizations of various degrees of specialization and permanence. The national associations collected scientists, supporters of science, and the interested public in large annual meetings that wooed inclusiveness by changing venue from year to year. The Gesellschaft deutscher Naturforscher und Ärzte was the oldest (1822), followed closely by the British

Association for the Advancement of Science (1831), and, later, by similar bodies in France, the United States, and Italy.

Internationalism was as distinctive a feature of the half-century before World War I as were the nationalisms it tried to ameliorate. Encouraged by cosmopolitan or humanitarian promptings, enabled by the efficient communication and transportation system of Europe, and supported by governments seeking international agreements in technical matters, international conferences on various branches of physics tripled, from 1.7 per year around 1875 to 5.5 per year around 1905. The logical apex of the international movement in physics was an organization to coordinate the world's work in physical science in accordance with a grand research plan. In 1912 Solvay partially filled this logical space by founding an Institut international de physique. Charles Edouard Guillaume, director of the Bureau international des poids et mesures (1875), then proposed the establishment of an international association of physical societies. That was in 1914. Soon the physicists Guillaume strove to unite, and those already brought together by Solvay, split into warring camps and prepared to mobilize their common science to their mutual destruction.

Fig. 23. A small piece of big physics. A section of the Tevatron tunnel at Fermi-lab. Protons accelerated in an evacuated circular pipe four miles in circum-ference attained a velocity 99.99995 percent that of light. The pipe resides within the lower line of boxed superconducting magnets. The upper line of magnets moved antiproton beams at room temperature.

6

FROM OLD WORLD
TO NEW

————•◦•————

Historians, philosophers, theologians, popularizers, and Nobel Prize committees concerned with twentieth-century physics tend to direct their attention to TOEs, or, more modestly, GUTs (grand unified theories of the four fundamental forces), and to the microphysics of cosmogony. Despite the bodily metaphors, the emphasis privileges the mind-expanding concepts of relativity theory, quantum physics, and the systematics of elementary particles. A secondary theme stresses the control of nature by physical theory as demonstrated by the release of nuclear energy. After this demonstration, great laboratories, national and international, employing machines as expensive as battleships, multiplied "elementary" particles as well as high-energy physicists and the army of technicians, scanners, computer specialists, engineers, grant officers, bookkeepers, secretaries, and contractors on whom TOE-and-GUT quests now depend. The symbol and acme of this enterprise in the short American century is Fermilab (see Figure 23).

Nevertheless, the romantic investigation of the smallest and the biggest things in the universe employs less than 20 percent of today's physicists if the divisional breakdown of the American Physical Society adequately represents the profession at large. The majority of physicists work on condensed matter, plasmas, computers, optics, lasers, polymers, fluid dynamics, materials, atomic, molecular, and nuclear problems, questions of military interest, and border subjects overlapping chemistry, earth sciences, biology, medicine, and engineering. They also culti-vate the latest sprigs of that old trunk of *physica, meteorologica*: climatology, plate tectonics, ionospherics, oceanography, and seismology.

During the interwar period, as the United States began to acquire its hegemony, concerns and problems originating in Europe ordered items of the agenda; after the war, American-ization gave precedence to different items and pursued them in different ways.

Legacies of World War I

While drifting westward toward the US, the center of gravity of physics experienced a tug from the Soviet Union and Japan. A few numbers will illustrate the vigor of expansion in previ-ously peripheral countries. The figure of merit is a doubling each decade between 1920 and 1940. That happened in the US for the annual harvest of PhDs in physics (50 to 200) and physicists in industry (400 to 1,600). The corresponding figures for Japan show a doubling time for PhDs of about seven years, for memberships in the Physical and Mathematical Society of Japan of 11 years (reaching 1,100 in 1940), and for research laboratories of 20 years (to 325). The rapid growth of physics in the Soviet Union may be indicated by an eight-fold increase in universities and colleges, a six-fold increase in scientific faculties and staff, and a 20-fold increase in scientific and technical employees of the Academy of Sciences.

The feedback mechanism for this explosive growth, and parallel but slower increases in European centers, was the discovery of the importance of physics for national defense, industrial development, and what Italian fascists and Japanese militarists called "autarky." The Great War had shown that determination and technology could make a nation independent of others for products and raw ingredients. Thus it was with nitrogenous fertilizers in Germany and with pharmaceuticals, fine chemicals, electrical appliances, and optical glass in the Entente powers. Among the first research establishments set up in Japan and the Soviet Union were optical institutes charged with perfecting glass for military purposes. In the late 1920s Italy followed suit with an institute for optics, one of the few successful state-supported high-tech research endeavors during Mussolini's regime.

Two other industries virtually created during the war, aeronautics and radio, gave employment to physicists in all developed countries after it. Government and industrial laboratories for radio and telephony multiplied, perfecting vacuum tubes, receivers, relays, and transmitters, promoting miniaturization and developing sound recording and playback. Physicists so employed helped to engineer several simultaneous revolutions by providing hardware for large-scale research, radios for domestic leisure, and the loudspeaker for politicians and propagandists. Fascists and Nazis battened on the big lie, the mass rally, and the amplifier.

The science and rhetoric of World War I still confront us in weather reports, which record the vision of Vilhelm Bjerknes (Bergen) and his school of climatologists. Cut off from meteorological information from the West during the war, they filled the hole with the theory of fronts between struggling air masses. They chose their terminology, and refined their concepts, with an eye to the shifting boundaries between the human masses then struggling in the trenches. A vast amount of data about the upper atmosphere, collected to guide the flights of planes and shells, stood ready to test and extend meteorological theories.

In addition to its legacy of hardware and practically minded scientists, World War I confronted demobilizing physicists with fundamental new research performed in sheltered niches during it. Einstein had pondered the curious fact that a body's resistance to motion has the same measure as its acceleration under gravity. To him the curiosity transmitted the truth that an observer experiencing motion when standing in a closed compartment would not be able to tell whether the compartment was accelerating upward or resting in a gravitational field. It followed that all observers can consider themselves at rest in their reference frames if they attribute any acceleration they feel to gravity. This equivalence permitted Einstein to geometrize gravitation 300 years after Galileo had intuited the possibility. The trick was to use non-Euclidean geometry. A wheel spinning at relativistic velocities has a different geometry than when resting because the circumference suffers a Lorentz contraction and the radius does not. Owing to the equivalence of acceleration (in this case of circular motion) and gravity, Einstein inferred that massive gravitating bodies warp the space around them so as to capture other bodies. Depending on the velocity and direction with which a celestial object approaches the warp, it can be swallowed like an apple, acquired as a planet, or deflected as a comet. Just like a Cartesian solar vortex! Einstein completed his geometrical gravity, "the general theory of relativity," in November 1915, the year in which his friend and colleague Fritz Haber enriched the art of war with poison gas.

According to general relativity, light should swerve when passing through warped space. Einstein suggested testing the prediction by comparing the direction of stellar rays that grazed the Sun during a total solar eclipse with their direction six months earlier. In 1918 American astronomers tried to detect the expected shift. They failed. A year later an English expedition found it. The confirmation aroused immense public interest in a world eager to raise its gaze to the stars. Subsequent measurements brought the Americans into agreement with the British and established Einstein's geometrical gravitation as a respectable oddity.

Bohr's quantum theory of the atom prospered mightily during the war. In neutral Denmark, he and his Dutch student, Hendrik Kramers, developed the correspondence principle, and in neutral Sweden Manne Siegbahn (Stockholm) and his students made precision measurements of characteristic high-frequency spectra of the elements that gave access to the deep structure of atoms. These measurements exploited two capital discoveries made just before the war. In 1911–12, Max von Laue (Munich), William Lawrence Bragg (Cambridge), and William Henry Bragg (Leeds) showed that X-rays can be refracted and reflected by crystals, which made possible X-ray spectroscopy and tipped the wave–particle duality toward waves; and in 1913–14 Henry Moseley (Manchester, then Oxford) inaugurated X-ray spectroscopy by detecting high-frequency line spectra characteristic of the elements from which they came. X-ray spectra from neighboring elements in the periodic table resemble one another more closely than optical spectra. Moseley found that their strongest line had a frequency that obeyed the simple Balmer-like formula, $\nu = K(Z-1)^2 \times (1-1/2^2)$. He interpreted the formula as a confirmation of the concept of atomic number and a signpost to the innermost region of the atom. The war prevented him from following it further.

Arnold Sommerfeld (Munich), who was too old to fight and impractical in war work, developed Bohr's theory to apply to complicated electron orbits by introducing additional quantum numbers. With them he accounted for X-ray and complex optical spectra. His systematic presentation of his postulates and results, *Atombau und Spektrallinien* (1919), gave demobilizing physicists the score of what he called the "atomic music of the spheres." Sommerfeld's music, though magnificent, was too formal for Bohr's taste as it introduced quantum numbers and ad hoc limitations on electronic transitions without much physical justification. Bohr preferred his method of intuition and conjecture, which only he, following what Einstein admired as his "unfailing tact," knew how to use.

The third fundamental wartime novelty in basic physics obtruded when Rutherford, snatching time from his war work, inquired how close to the nucleus Coulomb's law held. He found a few deviations of a peculiar kind in encounters between hard alpha rays and air molecules: instead of being repelled, the alpha disappeared and a faster, apparently lighter particle emerged. Rutherford supposed that he witnessed a game of marbles in which the alpha entered the nuclear precinct and knocked out a hydrogen nucleus (a "proton"), and conjectured that nuclei consisted of alpha particles, protons, and neutral particles each made up of a beta ray and a proton.

The legacy enriched

Atoms and electrons

In the summer of 1922 Bohr gave inspiring, perplexing lectures at Göttingen in which he assigned to every electron in the periodic table a pair of quantum numbers designating its energy and orbital angular momentum, and claimed that the correspondence principle determined how many of the elements so defined fell into each row of the periodic table. Among his auditors was Sommerfeld's prize student Wolfgang Pauli, who three years later assigned two more quantum numbers to each electron and employed an "exclusion principle" to derive the lengths of the rows of the periodic table, 2, 8, 18, 32. The principle—no two atomic electrons can have the same values of all four quantum numbers—was more admonition than cause. Its implication that antipathy reigned among electrons showed that some concepts of Greek *physica* remained useful; and its rapid acceptance measured how far atomic physicists had accustomed themselves to the irrationality of the microworld.

In the same year, 1925, another product of Bohr–Sommerfeld rearing, Werner Heisenberg, fresh from a stay with the strict mathematical theorist Max Born (Göttingen), played Bohr's guessing game well enough to find a calculus relating only observable features of spectral lines. The calculus presented frequencies and intensities in square arrays ("matrices") whose entries (the intensities) were located by the principal quantum numbers defining possible electronic transitions. Heisenberg, his former fellow student Pascual Jordan, and their teacher the formalist Born quickly sophisticated Heisenberg's insight so effectively that very few physicists could understand it. Fortunately for their sanity, Erwin Schrödinger (Zurich), a descendant of Boltzmann's school, found another calculus without recourse to matrices. Rather, he picked up a hint from the doctoral thesis of Louis de Broglie (Paris), who used relativity theory idiosyncratically to associate a standing wave with an electron moving in a stationary state. Schrödinger devised a differential equation for this wave (subsequently "the ψ-wave" or "wave function") that depended on the electron's energy. Calculators soon discovered that they could obtain the numbers in Heisenberg's matrices by easier and more familiar procedures using Schrödinger's waves.

The almost simultaneous discoveries of wave and matrix mechanics followed two different paths from the same demonstration: that an X-ray can strike an electron much as one billiard ball does another. This "Compton effect" (1922, after Arthur

Holly Compton, St. Louis) stampeded physicists into taking seriously the concept of the light particle ("photon") that Einstein had suggested as a "heuristic hypothesis" in 1905 to account for the photoelectric effect. De Broglie's coupling of a wave to a particle accepted the photon, whereas Bohr, who had rejected light quanta, tried to elude Compton's billiard-ball theory. After experiments decided for Compton, Bohr gave equal weight to wave and particle properties. Since a wave spreads energy whereas a particle localizes it, Bohr added the necessary premise that no realizable experiment can require a description employing these contradictory characteristics simultaneously ("complementarity," 1927).

Schrödinger interpreted the ψ-wave realistically, as a measure of the electrical density at every point in the atom. Many physicists hoped that he was right. The hope proved groundless and physicists accepted Born's elucidation that ψ^2 at any place gave the probability of finding the entire electron there. Heisenberg then drew from the mathematical formalism the conclusion that the microphysicist can never do better than reckon the probable, that not even a Demiurge could determine the exact position and corresponding momentum of an electron ("the uncertainty principle," 1927). To Einstein and other senior physicists this conclusion was a derogation of duty. If "'the old one'... is not playing at dice," neither should physicists. Several times Einstein constructed thought experiments that garnered more information than quantum mechanics allowed, and every time Bohr escaped by finding an inconsistency or omission in Einstein's examples. The defeat of his last attempt, made with two colleagues in 1935, persuaded almost all physicists who still took an interest in the argument that quantum mechanics does give a complete though probabilistic account of the phenomena of atomic and molecular physics.

By 1935, the quantum mechanics of electrons and radiation rested on a relativistic theory invented by Paul Dirac (Cambridge, 1928). Its results had been as surprising and gratifying as the radio waves unexpectedly generated by Maxwell's equations. Imposing the requirement of relativistic invariance made the electron's wave function into a four-component column matrix, which turned out to describe particles with a fourth degree of freedom and either positive or negative energy. Dirac interpreted the extra degree as an intrinsic (but non-mechanical) spin, thus rationalizing Pauli's quantum numbers and the concept of electron spin proposed to improve spectral systematics by Samuel Goudsmit and George Uhlenbeck (Leyden) in 1925. The negative-energy solutions appeared to describe a particle with positive charge. One that fitted the description (the "positron") quickly turned up in a context soon to be described. Meanwhile, several researchers, including J.J. Thomson's son George Paget Thomson (Aberdeen), detected the "de Broglie wave" of electrons. The subsequent award of a Nobel Prize to Thomson for showing that the electron is wavelike thus oddly complemented the prize his father had won for showing it to be particle-like.

Dirac's relativistic theory supposed that a high-energy photon could create an electron–positron pair, which, by subsequent mutual annihilation, could make another photon. Incorporating this demiurgic idea, theorists developed quantum electrodynamics, an account of electrons and photons real and virtual that predicted values of many measurable parameters to exquisite accuracy once they had developed rules to remove the infinities that pestered their calculations. That did not happen until after World War II.

Nuclei

The meager flux of high-energy alpha particles from natural radioactive sources did not score enough hits to explore nuclear architecture effectively. Several physicists proposed to increase the flux by accelerating beams of protons to the many hundreds of thousands keV (1,000 electron volts) deemed necessary to jam one into a light nucleus. Here the subtleties of quantum mechanics, brought to Rutherford's laboratory at Cambridge by the imaginative Russian physicist and humorist George Gamow, came to the aid of the marble shooters.

Gamow had noticed that the probabilistic ψ-calculus predicted that a proton could sometimes enter ("tunnel into") a nucleus with less energy than it would require if it behaved like a marble and the nucleus like a brick wall. Rutherford's lieutenant, John Cockcroft, calculated that 300 keV protons might tunnel through the barrier, and, in collaboration with Ernest Walton and the major electrical manufacturer Metropolitan Vickers, managed to reduce lithium atoms to alpha particles, the first artificial radioactive transformation. By 1932 they had taught their apparatus to hold 700 keV and disintegrated lithium in quantity.

Meanwhile, Ernest Lawrence (Berkeley) had devised a way to add energy in small steps to a spiraling charged particle so that the entire accelerating potential never had to be maintained on any part of the apparatus. The key to his "cyclotron" is that at nonrelativistic speeds the frequency with which a charged particle circulates when confined by a magnetic field does not depend on its energy. Lawrence and his students placed two hollow D-shaped electrodes within a squat cylinder with a proton source at its center, exhausted the space, applied an oscillating electric field across the gap between the "dees," slathered on sealing wax, shimmed the magnet, and, at the end of 1932, had a tiny beam at 1 MeV (million eV) (see Figure 24). That

Feb. 20, 1934.   E. O. LAWRENCE   1,948,384

METHOD AND APPARATUS FOR THE ACCELERATION OF IONS

Filed Jan. 26, 1932   2 Sheets-Sheet 1

Fig. 24. Patent on the cyclotron. The cyclotron principle—that the accelerated particles' angular frequency, $v/r = (e/m)H$, is independent of its energy at nonrelativistic speeds—allows the spiral motion to take place entirely under a constant magnetic field H. As shown on the right, the electric and magnetic fields automatically focus the particle beam.

same year James Chadwick (Cambridge) detected Rutherford's nuclear neutral particle ("neutron"). The simplest combination containing it, the deuteron, the nucleus of heavy hydrogen ("deuterium"), was isolated, again in 1932, a wonder-filled year for physicists if for no one else at the depth of the Great Depression.

As available energies increased, so did the number of elements transformed or disintegrated. The marble model persisted, however, in that experimenters supposed that the incoming and outgoing particles changed places immediately. But in 1934, Marie Curie's daughter Irène and her husband Frédéric Joliot (Paris) observed a delay; and soon nuclear physicists and chemists were making "activities" throughout the periodic table by shooting protons, deuterons, alpha particles, and neutrons at every element they could procure. The ever-enlarging cyclotron, with its high-power radio and vacuum technology, reached 60 inches and 220 tons in 1938. A year or so later it was producing prodigious currents of 16 MeV deuterons. Berkeley cyclotrons manufactured more activities, in quantity and variety, than all other laboratories, often laboring around the clock consuming charged particles and medical money to produce radioisotopes with desirable properties as tracers (notably carbon-14) and therapeutic agents (phosphorus-32). The cyclotron was preeminently, and, with its large attendant staff, capital costs, and operating expenses, symbolically American. Only eight existed outside the US at the outbreak of World War II, the largest, at Cambridge and Copenhagen, having 36-inch pole pieces. A man from Berkeley had to be summoned to get several of them started.

Just as the marble analogy blinded physicists to the phenomena of artificial activities, the anticipation that the descendants of a transformed element would lie close to it in the periodic table blinded them to the possibility of nuclear fission. Enrico Fermi and his group (Rome) saw and missed it in the course of irradiating uranium with neutrons. Lise Meitner and her chemist colleagues Otto Hahn and Fritz Strassmann (Berlin) persisted for years in identifying the products of uranium irradiation as "transuranic." Eventually, the chemical evidence persuaded Meitner's colleagues that the uranium nucleus could be split into two elements of moderate atomic weight. Meitner, then an émigrée in Siegbahn's laboratory in Stockholm, and her nephew Otto Frisch, then working in Bohr's institute in Copenhagen, proposed the mechanism of fission, which Bohr clarified and broadcast to the world in 1939. Everyone knowledgeable realized that the neutrons released by fission in a large sample of uranium might provoke a chain reaction, and a devastating explosion, before escaping from it.

Stars

After the excitement of 1920, interest in Einstein's gravitational theory flagged for a decade. His version described a stable, almost flat, closed universe—an updated Aristotelian cosmos. The equations also allowed expanding, contracting, and oscillating universes. The first to treat expansion as the true state of affairs was Georges Lemaître, a Belgian priest who had studied at Cambridge and Harvard, where he learned about Edwin Hubble's (Caltech) capital discovery that other "island universes" (galaxies) are receding from ours with speeds proportional to their distances from us. In 1931, Lemaître proposed that the expansion sweeping the galaxies along had a beginning, when the mass of the universe, encapsulated in a single atom, exploded. The steady-state Greek cosmos again gave way to a time-dependent Judeo-Christian world.

Just after the explosion, everything was radiation and maybe electrons. Where then did Thales' water come from? The question of the origin of the elements drew George Gamow, his Hungarian colleague Edward Teller (Washington, DC), and the world's leading nuclear theorist Hans Bethe (Cornell), all immigrants to the US, into the expanding universe. Gamow at first supposed that stellar interiors provided the ovens that cooked protons into the elements and that the tunneling mechanism he had discovered kept down the temperature required. In 1938 Bethe succeeded in defining the nuclear syntheses that power the stars, which did not require or reach a temperature at which nuclei heavier than helium's could be made. Gamow therefore relocated his ovens (unnecessarily, as it turned out) to the hotter regime of the early expanding universe.

In 1942, in the last of the Washington meetings organized by Gamow and Teller that had helped set the agenda of American theoretical physics in the 1930s, the participants agreed that, as stated in the report drawn up partly by Gamow, "the elements originated in a process of explosive character, which took place 'at the beginning of time' and resulted in the present expansion of the universe." Thus the physics of the laboratory, where the energies required to add a nucleon to a nucleus could be measured, made contact with speculations about the state of the universe at grossly inaccessible times and places. At this point

Gamow broke off speculating about the great explosion at the origin of time to help the US navy improve its gunnery in the here and now.

Other interwar business

Meteorologica

The discovery of radioactivity breathed new life into the old subject of atmospheric electricity created by Franklin's theory of lightning. Radioactive substances near the Earth's surface give off radiation that ionizes the lower atmosphere and keeps the Earth from cooling at the rate Lord Kelvin had calculated to deprive evolution of the time Darwin supposed necessary to create Kelvins. As expected, the ionizing radiation declined with height above the Earth's surface. Consequently the discovery that at higher altitudes ionization increases bewildered Victor Hess (Vienna, 1912) and other daredevil physicists who measured radiation from balloons. The old meteorological problem arose again: was the cause located beneath or above the Moon? Hess judged it to lie outside the solar system since it operated with equal strength day and night. After World War I, Millikan took up the question using war-surplus sounding balloons. In his opinion, incoming "cosmic rays" (as he called them) were photons created in space during the amalgamation of four protons. Absorption measurements and calculations of the energy released (via $E = mc^2$) at first favored Millikan's fantasy that cosmic rays were the "birth cries" of atoms. In 1929 Walther Bothe and Werner Kohlhörster (Berlin) found that most of the charged rays at sea level could penetrate ten inches of gold. Photons would have to have prodigious energies to create Compton electrons capable of such a performance. Bothe and Kohlhörster concluded that primary cosmic rays carry a charge.

The Earth itself provided the instrument needed to test this conclusion. Bruno Rossi (Florence), Lemaître, and Manuel Vallarta (MIT)—adapting Carl Størmer's (Oslo) calculations of the paths of particles causing the aurora borealis—showed that the Earth's magnetic field would bend the trajectories of incoming particles by an amount dependent on latitude. The earliest observers, including Millikan, found no such effect. But it exists, and is not small, some 14 percent between the magnetic equator and Chicago, as a worldwide survey led by Compton established in the early 1930s. His colleagues of various nationalities worked with standardized ponderous electroscopes encased in brass and lead supplied by the Carnegie Institution's Department of Terrestrial Magnetism.

The latitude survey counted all the cosmic rays that penetrated its instruments. Bothe and Kohlhörster had claimed to follow the path of one and the same ray through their gold brick. They had placed Geiger counters, which register the passage of charged particles, above and below the brick, and considered only events in which electroscopes connected to the counters deflected together. Their method excited Rossi and his fellow student in Florence, Giuseppe Occhialini, to do better. Exploiting vacuum tubes available for radios, they detected coincidences electronically and employed several counters simultaneously. Thus they discovered that a cosmic ray can sail through a meter or more of lead, implying energies of several GeV (1,000 MeV).

Electronic counters made possible efficient self-portraiture of a cosmic ray when arranged to trigger a cloud chamber just after the rays passed through it and them. The chamber, invented before the war by C.T.R. Wilson (Cambridge) and improved after it by P.M.S. Blackett (also Cambridge), records tracks by condensing water droplets around ions created by the passage of energetic charged particles. The intensity of ionization and the curvature of the tracks in the strong magnetic field maintained across the chamber gave information about the particles' charge, mass, momentum, and energy. Since the chamber must expand to make the droplets and remains sensitive for only a hundredth of a second, and since only one cosmic ray hits a square centimeter on earth per minute, the chance that a random expansion might catch a revealing track is small. Blackett's group inspected 400,000 tracks in 1924 to find six showing Rutherford's nuclear marble game. By hooking up Rossi's coincidence circuitry, brought to him by Occhialini, to the cloud chamber, Blackett got many fine pictures of the tracks of positive electrons. That was shortly after Carl Anderson (Caltech) chanced to find a beautiful example on one of the 3,000 exposures he made without coincidence circuitry.

The positron put in its appearance in 1932, the year of the neutron, deuteron, cyclotron, and Cockcroft–Walton. Five years later, Anderson and others at Caltech, Harvard, and Tokyo found the tracks of what they called "mesotrons" or "mesons"— particles weighing around 200 electron masses (200 m). The weight agreed with that of the hypothetical "Yukon," which Hideki Yukawa, the first important domestically trained Japanese theorist, had proposed in 1935 as the mediator of the "strong force" between neutrons and protons. The move had a precedent: physicists had already accepted a "weak force" to account for beta decay and an imponderable uncharged "neutrino" to conserve energy in the process. Naturally, they identified Anderson's mesotron with the Yukon.

The mesotron is incorrigibly unstable. Rossi, resettled in the US (1939), determined its half-life at rest by measuring the delay between its arrival on earth and the emission of the electron into which it decays. The result of two microseconds, obtained in 1943, was not long enough to bring mesotrons down through the atmosphere in the numbers observed at the surface. Since, however, they move at a speed comparable to that of light, they live detectably longer in motion than at rest, and so delivered the first direct confirmation of relativistic time dilation.

Coldness

Physicists gained the keys to another new kingdom at the same time they discovered cosmic rays. In 1911, Heike Kamerlingh Onnes (Leyden) and his associates, having liquefied helium in 1908, found that the electrical resistance of pure metals plummeted suddenly to zero at liquid-helium temperatures. This phenomenon, to which he gave the name "superconductivity," was not the greatest surprise he had in the kingdom of the cold. Liquid helium had an anomalously low density and would not freeze. These hints stimulated research after the war following a pattern similar to that of nuclear physics. A new-world laboratory challenging the old leader invented new costly apparatus that magnified effects a hundredfold. More laboratories entered the field and stranger phenomena came to light. Theorists competed to explain them with concepts only slightly less bizarre than the phenomena. And these concepts too followed a familiar developmental pattern. They arose from the application of quantum mechanics to fields founded in the decade before World War I.

The new-world site that challenged the old (in this case Leyden) was again Berkeley. The counterpart to Lawrence, William Francis Giauque, attained temperatures within a degree of absolute zero in 1933, after investing seven lean years in developing his technique of "adiabatic demagnetization." It requires a paramagnetic crystal cooled to liquid-helium temperatures where the little thermal energy it retains resides in the vibrations of its lattice and the disordered spins of its atoms. Giauque used a strong magnetic field to line up the spins and liquid helium to carry away the heat thus generated, as in the isothermal compression of a steam engine. Having insulated the specimen thermally, he turned off the field, the spins regained their disorder at the expense of the lattice vibrations, and the temperature fell, as in an engine's adiabatic expansion, to half a degree absolute. By 1938 he had arrived within 0.004 degrees centigrade of absolute zero (0.004 K). Experimentalists thus acquired a tool for studying superconductivity and other odd phenomena in regions well below 4.2 K, the liquefaction point of helium. The odd phenomena, which included a sudden, million-fold jump in heat conductivity and an abrupt vanishing of viscosity in helium, gave rise to the concept of superfluidity (1938).

Effective theories of superconductivity and superfluidity drew on a quantum reformulation of the old electron theory of metals synthesized by Sommerfeld and Bethe in 1933. Building on the insights of Heisenberg's student Felix Bloch (Leipzig, then Stanford), they treated unbound electrons in crystals at low temperatures as free except for the exclusion principle, which forces them to fill up the lowest quantum states available. Only those near the top of this electron sea can take up additional energy as the temperature rises and so sail along it or above it; in a perfect solid, each such electron goes on its separate way subject to the same ionic forces. The new model agreed with the old theory where it had worked (as in the Wiedemann–Franz law), and explained its failures (as in specific heats) where it had not. These results, obtained by European theorists, pertained to ideal solids. Beginning in 1933, Eugene Wigner and his student Frederick Seitz (Princeton) began the hard task of calculating what Pauli called the "dirty effects" of impurities, and other American groups, notably John Slater's at MIT, soon mucked in.

The one-electron model did not account for superconductivity. That became more embarrassing as information about it mounted during the 1930s. There were two main centers of research: Oxford, where Franz Simon, Fritz London, and other German-Jewish refugee physicists, descended intellectually from Nernst and, in Simon's case, Giauque, set up a thriving low-temperature laboratory with the help of British industry; and Moscow, where Peter Kapitza, a former associate of Rutherford, and the almost legendary theorist Lev Landau, worked in the institute given to Kapitza as consolation for his detention in the Soviet Union against his will. Although some of these men glimpsed the nature of a workable theory of superconductivity, they kindly allowed the Americans to invent it.

Americanization

More war spoils

During World War II, physical science and engineering created radar, which led to microwave technologies and the laser; atomic bombs, which brought nuclear power; the V-2 rocket, which launched the aerospace industry; and the "bombe," the essential device for breaking the Enigma codes, which, with electronic analyzers constructed for calculating ballistic trajectories, pioneered the electronic computer. This intrusive device emerged in the late 1940s in materializations of Alan Turing's conception of a "universal machine," which related to the computer as Carnot's theory did to the steam engine. The ENIAC, the first postwar electronic computer, was rich in vacuum tubes (some 17,468), poor in memory, and programmable only by changing its wiring. Developed at the University of Pennsylvania under an army contract to calculate firing tables, it soon switched to working on the hydrogen bomb. Its successor, the MANIAC, built on the architecture created primarily by John von Neumann and demonstrated on the machine he designed at the Princeton Institute for Advanced Study, contained its own instructions and could be reprogrammed without rewiring. Von Neumann architecture has survived throughout the revolutionary advances made possible by transistors, microprocessors, and gigabyte memories.

As the only major belligerent to emerge from the war rich and relatively unscathed, the US was best positioned to capitalize on the computer and other inventions. To augment its head start, it sent scavengers behind its advancing armies to seize information about developments in Germany. In some cases, as in the transfer of Werner von Braun's rockets and technicians to Alabama ("Operation Paperclip"), the US picked up men as well as hardware without inquiring zealously about their degrees of Nazism. The French, British, and Russians also took what they could, but they did not match the Americans in their ability to exploit their conquests.

Americanization meant not only an emphasis on big machines and large collaborations, and hurried competition fostered by ambition and federal funding agencies, but also a neglect of matters that the generation of Bohr and Einstein and their disciples had thought important. Just as Newton's followers in the eighteenth century dropped his scruples about distance forces and ignored his rules of philosophizing, so postwar Americans did not bother themselves with the enigmas of quantum physics or, in general, with the high culture of the prewar Europeans. The uncomplicated unlearnedness of Lawrence and his associates was closer to the norm than the tortured culture of Robert Oppenheimer and the young theorists who fashioned themselves after him.

The Americanization of theoretical physics made its postwar debut in a conference held in June 1947 at Shelter Island to insulate the participants from the demands society had placed on "brilliant young" physicists—the only type, apart from Einstein, it knew. The planners consulted Pauli, then at Princeton, and ignored his suggestion of a large conclave reconvening the creators of quantum mechanics. They opted for a small specialist meeting featuring upcoming Americans, homegrown theorists from Oppenheimer's school at Berkeley and Caltech and other youths whose science the war had honed. The plan of the meeting, drawn up by John Wheeler (Princeton), reads like the agenda for a fast-paced attack on a wartime problem. The brilliant young things were to begin with the question whether a logical, consistent, and comprehensive quantum theory exists; if yes, to frame it; if no, to define the limitations of existing theory and propose ways to extend it. Three days later the participants were to draft a collective statement on one page setting out their answers to the initial question.

Two-thirds of the participants were American by birth and most of the others were émigrés of the 1930s. Bohr's former collaborator Kramers (Leyden) urged in vain for consideration of Copenhagen's old-world view that the troubles of quantum electrodynamics ("QED") could not be overcome without a better classical electron theory on which to apply the correspondence principle. The intellectual leader of the meeting was Oppenheimer. Performing as he had when director of Los Alamos, he went straight to current problems like the demoralizing infinities of QED and the puzzling behavior of the mesotron. Its short life was too long for a Yukon, which had to interact strongly with matter. Robert Marshak (Rochester) made the generous suggestion, soon confirmed, that the Yukon was not the only meson in the world. As for the infinities, pragmatic "renormalization" of QED by Richard Feynman and Julian Schwinger (and independently by Yukawa's classmate Sin-Itiro Tomanago) saved the phenomena to as many places of decimals as anyone could measure.

A second conference, held in 1948 with most of the same participants but with Bohr in place of Kramers, inspired an instructive clash between new and old. Feynman (Cornell) sketched his version of QED with diagrams of the sort that soon became indispensable. Bohr did not like them: they violated the uncertainty principle by assigning definite paths to particles and they represented positrons incomprehensibly as electrons moving backward in time. No matter. Feynman and Schwinger (Harvard) were the new world-class theorists. Freeman Dyson, an Englishman who made his career at Princeton, observed this coming of age: "In the public mind, experimental science is a thing only Americans know how to do ... [T]he new Schwinger theory tended to be acclaimed as a demonstration that even in theoretical physics America has nothing to learn."

The US had learned to produce not only theorists of the Jewish-cosmopolitan-extrovert type like Oppenheimer, Feynman, and Schwinger, but also of an introverted, unassuming, gentile type. John Bardeen began neither as a physicist nor as a New Yorker, but as a petroleum engineer in the Midwest. Sickening of oil, he enrolled in graduate school in Princeton and worked on solids under Wigner. He began his career as a physicist at Bell Labs where, at the end of the war, he joined a small group under William Shockley (a student of Slater's from MIT) dedicated to making an amplifier using semi-conductors like silicon and germanium, which had been studied minutely for use in radar detectors. Bardeen's mastery of the quantum theory of solids together with the experimental inventiveness of Walter Brattain and the support of Shockley resulted, in 1947, in the invention of the transistor and a revolution in electronics. While Bardeen was in Stockholm in 1956 to collect the Nobel Prize awarded to him, Brattain, and Shockley, his theorist colleagues Robert Schrieffer and Leon Cooper (all then at the

University of Illinois) continued work on his idea that interactions between vibrations of the crystal lattice and electrons
skimming the surface of Sommerfeld's electronic sea supplied
the mechanism of superconductivity. The model was soon perfected. It was back to Stockholm for Bardeen, the only two-time
winner of the Nobel Prize in Physics.

Nuclear inheritance

In 1944, drawing up plans for the postwar era, Lawrence reck-
oned that his laboratory would require around $85,000 annually
and gifts in kind from the army's Manhattan Engineer District to
do well in peacetime. A year later, knowing that the bomb
project would succeed and that a worried government would
continue high-level investment in nuclear physics, he raised his
expectations a hundredfold. In the last days of the District (it
became the civilian Atomic Energy Commission, or AEC, on
New Year's Day, 1947), Lawrence obtained hundreds of kilo-
dollars' worth of radar tubes and other electronics for incorporation into the three major accelerators his laboratory had in
hand. One of these, the 184-inch synchrocyclotron, started up
just before the AEC did, with the presumed capacity of making
mesons. Attempts to capture them on photographic emulsions
failed, even after a cosmic-ray group led by Cecil Powell (Bristol)
had confirmed Marshak's conjecture by photographing the birth
of a Yukon (see Figure 25). Its parent, the "pi-meson" or "pion," is
the mediator of the strong force. Giulio Lattes (Brazil), who had
worked with Powell, came to Berkeley and showed the cyclo-
troneers how to photograph pions. As in induced radioactivity
and nuclear fission, a Berkeley accelerator had produced the

Fig. 25. Birth and death in the microworld. An emulsion stack exposed to cosmic rays gave the first proof of the two-meson hypothesis. In the example furthest left, a pion on the right strikes a molecule, giving rise to a muon that proceeds to the left until it decays into an electron and a neutrino.

means but not the men to make a discovery realized in Europe with much simpler equipment. However, the future in corralling specimens for what Americans called "the particle zoo" thereafter belonged to the big machine.

The AEC, uncertain whether better acquaintance with the pion might not be necessary for the next generation of nuclear bombs or power, authorized a machine in the 2–3 GeV range for its laboratory at Brookhaven on Long Island. Berkeley kept ahead with a 6 GeV version expected to produce antiprotons. The huge energies involved required a new design, in which protons could be accelerated to relativistic speeds not by spiraling out under the pole pieces of a large magnet, but by being confined by small fast-acting magnets in a narrow pipe laid out in a circle over 36 meters in diameter.

When completed in 1954, the "Bevatron" was an engineering marvel that could accelerate protons through four million turns (300,000 miles) in 1.85 seconds to 6.2 GeV. The following year, using an elaborate "telescope" of magnets and counters, a group led by Fermi's associate Emilio Segrè detected the antiproton. The detection was noteworthy not only for confirming the general theory of antimatter (and the applicability of Dirac's equation to protons) but also for revealing serious difficulties in apportioning credit for discoveries in big-machine physics. The physicist-builders of the Bevatron certainly deserved some credit. The designer of the quadrupole focusing magnets in the telescope, Oreste Piccione, an Italian émigré working at Brookhaven, demanded it. But only Segrè and his senior collaborator Owen Chamberlain received the Nobel Prize. Piccione's subsequent suit for a share was thrown out of court not because it lacked merit but because he waited too long to bring it. Hundreds of physicists, from graduate students through veteran scientists, may now work on a single "experiment." The feeling of being another cog in a great machine, and the certainty that the project leaders would get the credit for its results, drove many individualists into other sorts of work.

Nonetheless, accelerator laboratories multiplied. While the Bevatron settled down to making K mesons ("kaons") and other strange particles, and its attendants developed the bubble chamber—a detector that rivaled the main machine in cost and complexity and depended on cryogenics developed for the hydrogen bomb—competitors first nationally and then internationally came to threaten Berkeley's lead in high-energy physics (HEP). Fermilab, founded near Chicago in 1967, took the lead. Berkeley compensated by diversifying into projects of more immediate public utility—in materials science, energy, and geophysics—and by cooperating more fully with other laboratories. In this way it copied Europeans who, being unable after the war to match US investments in HEP, banded together to form the Centre Européen pour la Recherche Nucléaire (CERN), which now has the largest accelerator in the world, whose circumference of 27 km crisscrosses the French-Swiss border near Geneva (see Figure 26).

While pioneering in international scientific collaboration, CERN has been a cornucopia of elementary particles. Most recently, in 2012, it had the highly hyped honor of detecting the heavy particle postulated by Peter Higgs (Edinburgh) in the 1960s to account for the mass of other particles. In the interim, the keepers of the particle zoo distributed the animals into

Fig. 26. Very big science. A view of CERN indicating the route of the Large Hadron Collider under France and Switzerland. In 2009 its greater capabilities caused the Tevatron to shut down. Other acronyms (ALICE, ATLAS) designate experimental sites; SPS (Super Proton Synchrotron), CERN's first big instrument, now supplies the main ring with its initial beam.

families and tribes, and reduced nucleons and other heavy particles to combinations of "quarks" held so firmly together by "gluons" that experiment can never pry them apart. According to this standard model, whose playful American terminology contrasts pleasantly with Whewell's classical coinage, the nuclear, weak, and electromagnetic forces were one and the same during the very early universe. This GUT is not yet a TOE, as gravity remains outside it.

Radar spin-off

Immediately after World War II, English scientists who had worked on radar copied Galileo and directed their war-surplus equipment toward the heavens. The first important groups grew in Cambridge (J.A. Ratcliffe, Martin Ryle) and Manchester–Jodrell Bank (Bernard Lovell). They mapped radio sources in the solar system and identified many in other galaxies. In one source Maarten Schmidt (Caltech) found evidence of a red shift implying great distance, which, with its observed brightness, indicated an output of energy some hundreds of times that of our entire galaxy. Many such "quasi-stellar objects" (quasars) then put in an appearance, as did "pulsars" (Jocelyn Bell et al., 1967), which emit staggering bursts of energy at regular short intervals. These objects, and other odd stellar species uncovered by X-ray, infrared, radio, and visual observations, helped rekindle interest in cosmology.

The US, though the major wartime power in radar, was slow to take up radio astronomy, but when it entered around 1950 it possessed the first large telescope built to detect centimetric microwaves. In this region, in 1965, two physicists at Bell Telephone Laboratories discovered an annoying universal background radiation while surveying sources of noise disturbing radio communications. With the help of Robert Dicke and others at Princeton who had revived Gamow's cosmological ideas without knowing about them, the Bell discoverers, Arno Penzias and Robert Wilson, identified the ubiquitous noise with radiation left over from the aboriginal explosion (or "Big Bang") of the universe. According to Gamow and his student Ralph Alpher, radiation had gone its own way, disjoined from matter, shortly after the Bang and subsequently cooled as it expanded with the universe. The noise that bothered Bell Telephone had a temperature of about 3 K, not far from that expected from the expansion rate and Planck's formula. The community of cosmologists was then enjoying a dispute between Big Bangers (Gamow and the Princeton group), and Steady-State Men, who supposed that spontaneous creation of hydrogen atoms compensated for the dilution effected by expansion (Fred Hoyle, Hermann Bondi, and Thomas Gold, all Cambridge). The radiation at 3 K settled the question for all but a few diehards. In 1992 George Smoot (Berkeley) announced that his team had found a small inhomogeneity in the background radiation, which echoes irregularities in the primitive universe that once seeded stars and galaxies.

Radar expertise gave birth to a discovery of greater importance to most people than cosmic noise. In 1954 Charles Townes, who worked at Bell Labs from 1939 to 1947 on centimetric radar, combined war surplus with Einstein's old idea of "stimulated emission," a quantum process by which incoming light entices an atom in an excited state to radiate. The resulting device, the maser, which two Nobel laureates at the university (Columbia) to which Townes had moved from Bell Labs, told him would not work, excited molecules. Townes and others soon found ways to promote the maser to optical frequencies. The uses of the new device, the laser, in physics and everyday life are legion: holography, rock concerts, surgery, light art, CD players, barcode readers, and so on. The *Handbook of laser technology* published 50 years after Townes' invention of the maser required 2,665 pages to describe them all.

Earth science

Steering missiles, rockets, space probes, and nuclear submarines, testing nuclear explosives above and under ground and water, monitoring test-ban treaties, and searching for uranium and other strategic materials required exact information about the earth, ocean, and atmosphere. The relevant specialties—seismology, ionospherics, meteorology, and oceanography—had received little funding or prestige between the wars in comparison with nuclear physics. After World War II the US government threw money at them in order to perfect its knowledge of hostile environments, and joined enthusiastically in underwriting the multi-billion-dollar International Geophysical Year dedicated to the worldwide collection of pertinent data. At the same time, 1957, the earth sciences received a pleasant jolt from the launch of Sputnik. The National Aeronautics and Space Administration (NASA), set up to close the "space gap" and to put a man on the Moon, needed to know what its rockets, satellites, and space vehicles would encounter en route, in orbit, and on arrival. Bits of moon rock, displayed around the US, promoted earth scientists as the mushroom cloud had nuclear physicists.

Closer to Earth, flights of meteors, the solar wind, and the Earth's magnetosphere have become as predictable as the tides and even the weather, which is now forecast more accurately by meteorologists than by astrologers. This improvement was another dividend from military investments. Before World War II, many meteorologists rated the concept of fronts as "risky" and the prospect of a dynamical weather theory as "unpromising." After the war, Jule Charney (Princeton, MIT) and von Neumann adapted geostrophic models (in which winds blow along isobars) to computation by electronic computers. Models incorporating barocline instabilities (mismatches between the gradients of temperature and pressure) then developed together with computer capability and data collection via radar and satellite. These conceptual and computational advances have benefited meteorologists, other earth scientists, and also, by helping to identify anthropogenic contributions to climate change, the rest of the human race.

The stimulation of interest in earth sciences by government investment after World War II led to a fundamental discovery. Oceanographers detected that seafloor sediments and high ridges lying in the middle of the great oceans were relatively youthful, and that the foothills of the immense mid-ocean mountain chains enjoyed a surprisingly benign temperature. Further investigation disclosed that the temperate regime arose from melted rocks bubbling up from cracks in the ocean floor. In 1960 Henry Hess (Princeton) conjectured that the warmth and youth of the sediments indicated that molten magma rose from the Earth's interior at these ridges and spread the seafloor apart. That could not continue very long unless the spreading floor disappeared somewhere. Obvious loci for such brutal events are fault lines showing volcanic activity.

The developers of this theory had in mind the conjecture made in 1912 by Alfred Wegener (Marburg). Arguing from the distribution of species, the shape of continents, and the presence of coastal mountain ranges, he imagined that the Earth's entire landmass once fitted together. "Pangaea" subsequently broke apart and the continents drifted to where we find them. Geo-physicists persuaded by Kelvin's arguments for a rigid Earth rejected itinerant landmasses. In the modern version of Wege-ner's theory, it is not the continents, however, but "plates" on which they and the oceans ride, that migrate. This theory, "plate tectonics," received timely support from an entirely disparate line of inquiry into the residual magnetism of rocks. Geologists studying the meanderings of the Earth's magnetic poles as recorded in ordinary rocks reached the astonishing conclusion that the Earth's field has reversed from time to time, the south pole going north and the north south. The most recent of these reversals took place less than a million years ago.

In 1963 Fred Vine and Drummond Matthews (Cambridge) and Lawrence Morley (Canadian Geological Survey) suggested that the sea floor had recorded the magnetic reversals in stripes parallel to the vents from which the molten rock emerges. And this is what the oceanographic research ship *Eltanin*, operated by Columbia's Lamont Geophysical Laboratory, discovered. The measurements, which disclosed clean symmetrical reversals marching out to 2,000 km, were all the more persuasive because the measurers did not have the Vine–Mathews–Morley hypo-thesis in mind. The travels of the continents, as revealed with the help of sensitive magnetometers, mass spectrometers, argon-potassium radioactive dating, and other instrumentation derived from military technology take place at the brisk tempo of 4.5 cm per year.

The resources of some international facilities—like CERN and the European Space Agency, high-profile national ones like Russia's space program, and some industrial research laboratories—have long challenged American supremacy in physics. More recently, China and India have added their appreciable weight to the competition. Nonetheless, the US still dominates research. This is not merely because of the amount of money that has poured into its universities and national laboratories from federal and state government, the military, and industry. Indeed, its rate of financial investment in science and technology falls below the trend line in the developed countries of 3 percent of GDP. A very rough indicator of the continuing dominance of the US is its harvest of Nobel Prizes in physics, which, during the last 60 years, has been equal to or greater than the number awarded to the rest of the world collectively. American dominance will probably continue for a time even as international pressures progressively reduce it. US research universities train foreign graduate students and postdocs as well as domestic ones, and do so efficiently in the approximation to English that is the international language of science.

7

THE QUINTESSENTIAL

F or 20 centuries and more deduction of the place of God and man in the universe was an accepted, even an expected, activity of *physici*. Perhaps the last significant expression of this corporate responsibility occurred in England around the middle of the nineteenth century in eight volumes commissioned by the Royal Society of London under a bequest from the last Earl of Bridgewater. The volume demonstrating (as Bridgewater's bequest directed "the power, wisdom, and goodness of God, as manifested in the creation") via physics was the work of the same William Whewell who invented the term "physicist." Many individual physicists since have tried to unite traditional belief with their science, for example, Pascual Jordan, who found room for the freedom of the will in Bohr's interpretation of quantum mechanics.

Bohr applied complementarity in the Greek direction, obtaining insight into the human condition from his understanding of physics. According to him, just as the physicist cannot give a complete description of his experiments without invoking formally contradictory concepts, like wave and particle, so the philosopher must rely on opposing concepts, like free will and determinism, for an exhaustive description of the world. And just as no realizable experimental arrangement can require a quantum entity to behave simultaneously as a wave and a particle, we cannot experience freedom and determinism together. Freedom applies in prospect, determinism in retrospect, religion not at all. Bohr once intended to write a book against religion to fulfill the moral obligation that his understanding of physics placed on him. Similarly, Einstein's constant challenge to Bohr's "tranquilizing philosophy or religion" responded to a moral imperative. It was not a reactionary objection to novelty, but a duty to science, to try to maintain the ideal of space-time description. His God who does not play dice represented this ideal: Einstein rejected the concept of a personal deity and an afterlife as firmly as Bohr did.

The modern evolution of the physicists' world picture continues the ancient program of locating human beings in a directionless universe. We learn that the explosion that created it had to be nicely balanced to produce a state suitable for life. If too forceful, the primitive ingredients would have been scattered too far to form stars and galaxies; if too weak, the universe would have lasted only a second or two before sinking back into nothing. Apparently the initial power exercised, and the strengths of the forces subsequently at work, supplied the time needed (some 15 billion years) to develop intelligent life, which now uses its own existence as a criterion in designing accounts of the cosmos. This "anthropic hypothesis," so named by

Brandon Carter (Cambridge) in 1970, tickled the fancies of strong speculative minds, gave comfort to believers who saw the hand of God in the fine-tuning of the initial conditions, and appealed to atheists who interpreted our universe as a statistically improbable case of all possible universes originating in big bangs.

Another inference, implying a Bungling Demiurge rather than an Intelligent Creator, comes from the collection of investigations known as SETI (search for extraterrestrial intelligence). The US Congress voted a substantial sum to look for evidence of civilizations capable of producing electromagnetic messages. Not a peep has been heard despite the plausible assumption that among the billions and billions of possible habitats, some must have grown other beings during the last 15 billion years capable of inventing television. Is the universe not isotropic? Why then the silence? Perhaps the advanced civilizations we might imagine have had the foresight and skill to cordon off our corner of the universe. Or, perhaps more likely, electromagnetic civilizations do not survive long enough to give us a chance to tune into ET TV.

The universe as interpreted by modern physical science is horrifying as well as inspiring whether we are alone in it or not. The Earth, with its crashing plates, earthquakes, eruptions, hurricanes, and tsunamis is constantly exposed to bombardment from large meteors capable of wiping out most advanced life forms, as happened here only 66 million years ago. If a meteor or comet does not get us, we are sure to be destroyed by our own Sun. The heavens, which Aristotle pictured as a domain of eternal serenity, turn out to be filled with suicidal stars, black holes, super-energetic quasars, forbidding temperatures, and deadly radiations. Physicists believe that the universe originated in a catastrophic explosion of unimaginable violence owing to a chance fluctuation of nothing at all that scattered stars in their billions, and dark matter in its veil, through an indefinitely expanding or pulsating space. What, to repeat the quintessential question, is the place of humankind in this awesome world?

Steven Weinberg, who won a Nobel Prize for his part in advancing a TOE, gives the obvious adult answer: "The more we know about the universe, the more it is evident that it is pointless and meaningless." That is no reason not to try to comprehend it, however. Weinberg again: "The effort to understand the universe is one of the very few things that lifts human life above the level of farce, and gives it some of the grace of tragedy." In this telling, the purpose of the gigantic, finely tuned experiment is to give life the time and environment to evolve into physicists capable of developing theories of everything and inventing electromagnetic civilizations incapable of survival.

While awaiting the end, physicists pursuing modern versions of Greek speculations have exciting problems to solve. David Gross (Santa Barbara) has made a list of them. Here is the first: "can we actually say what happened at the beginning of the universe?" Hardly. We cannot even say what is happening now, since we are largely ignorant of the nature of the dark matter supposed to make up most of the mass of the universe. Perhaps CERN's upgraded Large Hadron Collider, coming back into play in 2015, will make gluinos and neutralinos, the superparticles that supersymmetry offers as candidate constituents of the dark filling of the heavens. Physicists who remember Aristotle call this filling "quintessence." Although gravity still eludes efforts to unite it with the three other fundamental forces, the case may be no worse than dark matter. And the worldwide collaboration Advanced LIGO (Laser Interferometer Gravitational-Wave Observatory), also to go on line in 2015, may detect the gravity waves that general relativity predicts as a product of collisions between black holes.

Why (Gross continues) do the fundamental parameters of the physicists' world picture have the values we find? No Pythagoras has arisen to explain why the top quark (whose existence was established at Fermilab after the inspection of 500 trillion collisions) weighs 100,000 times as much as the up quark, or the proton almost 2,000 times as much as the electron, or why nature has been so generous as to provide physicists with three families each of quarks and leptons (electrons, light mesons, neutrinos). Why is the proton stable? Why does the universe contain so little antimatter? Are the universal constants really constant? Is the concept of space-time fundamental? Does the latest fad—string theory—have the answers? Perhaps. But, alas, Gross writes, "we really do not understand what string theory is."

Is there time to learn?

Dedeifying and deanthropomorphizing nature produced physics, and physics now claims control of large swathes of territory previously ruled by capricious deities and organized religion. If the consequent picture of nature alarms humankind, that might reflect the healthy recognition that the world was not designed for us. The discoveries that no power made the universe for our pleasure, exploitation, or fright, and that there is no divine plot to history, can be liberating for people with the courage to accept them. Physics has given civilization a somber, disturbing, and challenging world picture, many fertile and some terrifying inventions, and notice of responsibility for the outcome of the human story. If humankind accepts the responsibility and the concomitant loss of providential deities and sacred dicta, the human species might beat the odds against the survival of an electromagnetic civilization, preserve the Earth, and, in the fullness of time, arrive at several satisfactory theories of everything.

REFERENCES

INTRODUCTION

1 Glashow, Sheldon, and Leon Lederman. "The SSC: a machine for the nineties." *Physics today*, 38:3 (1985), 28–37.

CHAPTER 1

9 "the source or cause": Aristotle, *Phys.*, 192b22. In *The Works of Aristotle translated into English*. Ed. W.D. Ross. 12 vols. Oxford: Oxford University Press, 1910–52.

9 "Of all who have discussed": Aristotle, ibid., *Metaphys.*, 988a.

10 "rule [that] applies": Aristotle, ibid., *Phys.*, 198a18.

11 "Same Stuff": Aristotle, *Meteor.*, in works, ed. Ross, 370a25–6.

14 "It is fitting": Francis M. Cornford, *Plato's cosmology: the Timaeus of Plato*. Indianapolis: Bobbs-Merrill, n.d. (The Library of Liberal Arts, 101.), quoting *Timaeus*, 29D.

16 "The great world's origin": Ovid, *Metamorphoses*. Tr. A.D. Melville. Oxford: Oxford University Press, 2008, p. 366.

17 "Many philosophical lineages": Seneca, *Natural questions*. Tr. Harry M. Hine. Chicago: Chicago University Press, 2010, p. 135.

18–19 "[T]he rapidity of its motion" and "the position of earth": Plutarch, *De facie quae in orbe lunae apparet*. Ed. Harold

Cherniss. In Plutarch *Moralia*, XII. Cambridge: Harvard University Press, 1995, pp. 59, 75.

21 "mere accessories" and "every act": quoted in G.E.R. Lloyd, *Ancient culture and society*. 2 vols. Cambridge: Chatto & Windus, 1970–3, 2, pp. 92, 94.

21 "corruptors and destroyers": Ibid., p. 94.

21 "courteous, just, and honest," "principles of physics," and "dig down": Vitruvius, *The ten books on architecture*. Tr. Morris Hicky Morgan. Cambridge: Harvard University Press, 1914, p. 21.

26 "strange and wonderful": pseudo-Aristotle, *Mech.*, in Aristotle, *Works*, ed. Ross, 855b25–30.

27 "There was never any": Pliny the Elder, *The historie of the world: commonly called the naturall historie*. Tr. Philemon Holland. 2 vols. London: A. Islip, 1635, 2, p. 586.

30 "disengaged from": Macrobius, *Commentary on the dream of Scipio*. Tr. W.H. Stahl. New York: Columbia University Press, 1952, p. 171.

31 "this universe": W.H. Stahl, Richard Johnson, and E. L. Burge, *Martianus Capella and the seven liberal arts*. 2 vols. New York: Columbia University Press, 1971–7, 1, p. 21.

32 "circling by their own surging": Ibid., 1, pp. 176–7, 2, pp. 317–19.

CHAPTER 2

43–4 Ibn Tufayl, Muhammad. *The history of Hai Eb'n Yockdan, an Indian prince: or, the self-taught philosopher*. London: R. Chiswell, 1686.

44 "I evermore": Omar Khayyam, *Rubáiát*, verse 27. Tr. Edward Fitzgerald. Ed. Daniel Karlin. Oxford: Oxford University Press, 2009, p. 29.

46 "observation of the stars": Régis Morelon, "Panorama général de l'histoire de l'astronomie arabe." In Rashdi Rashed, ed., *Histoire des sciences arabes*. 3 vols. Paris: Seuil, 1997, 1, p. 30.

48 "like snow upon the desert": Khayyam, *Rubáiát*, verse 14. Ed. Karlin, p. 23.

54–5 *The Qur'an*. Tr. M.A.S. Abdel Haleem. Oxford: Oxford University Press, 2008.

60 "their instruments were more precise": Hakim Mohammed Said and Ansar Zahid Khan, *Al Bīrunī: his times, life, and works*. Karachi: Hamdard Foundation, 1981, p. 70.

60 "the most important document": Ahmad Y. Al-Hassan and Donald R. Hill, *Islamic technology: an illustrated history*. Paris: UNESCO, 1986, pp. 58–9.

CHAPTER 3

66 Tannery, Paul. "Une correspondance d'écolatres du xi. siècle." In Tannery. *Mémoires scientifiques*. 17 vols. Toulose: E. Privat, 1912–50, 5, pp. 103–11.

66 "the ocean tide": quoted from Pierre Duhem, *Le système du monde: histoire des doctrines cosmologiques de Platon à Copernic*. 10 vols. Paris: Hermann, 1954–9, 3, pp. 16–20; "in 800 years": C.C. Gillispie et al., eds., *Dictionary of scientific biography*. 18 vols. New York: Scribners, 1970–81, 1, p. 565.

67 "the Arabic pomposity": Dorothea Metlitzki, *The matter of Araby in Medieval England*. New Haven: Yale University Press, 1977, p. 5, quoting the Bishop of Cordoba.

68 "our civilization": Ibid., p. 11, quoting Robert of Morley.

68 "A little learning": Alexander Pope, "Essay on criticism," 2.15, in *Collected poems*. Ed. Bonamy Dobrée. London: Dent, 1965, p. 122.

69 "knowing nothing": "Epistola Pharamellae," in Roger de Hoveden, *Chronica*. Ed. William Stubbs. 4 vols. London: Longman et al., 1868–71, 2, pp. 296–8.

78 "wholly unknown": quoted from Bacon's *Opus maius* by A. C. Crombie and John North, in Gillispie et al., *Dictionary of scientific biography*, 1, p. 380.

78 "It shall come to pass": *Genesis*, 8: 13, 14, 16.

81 "we are brought near" and "angelic art": quoted from Noel Swerdlow, "Science and humanism in the Renaissance: Regiomontanus' Oration on the dignity and utility of the mathematical sciences." In Paul Harwich, ed., *World changes: Thomas Kuhn and the nature of science*. Cambridge: MIT Press, 1993, pp. 133, 165.

81 "man is a *magnum miraculum*": quoted from Frances A. Yates, *Giordano Bruno and the Hermetic tradition*. Chicago: Chicago University Press, 1964, p. 35.

90 "intoxicated, crazy" and "novel, unheard-of": William Gilbert, *De magnete*. Tr. P.M. Mottelay. New York: Dover, 1958, pp. xlviii, xlix.

CHAPTER 4

95 "New philosophy calls all in doubt": Donne, quoted from Marjorie Nicholson, *Science and imagination*. Ithaca: Cornell University Press, 1956, p. 53.

97 Bacon, Francis. *The major works*. Ed. Brian Vickers. Oxford: Oxford University Press, 2002.

98–9 Descartes, René. *Discourse on method, optics, geometry, and meteorology*. Tr. Paul J. Olscamp. Indianapolis: Bobbs-Merrill, 1965. (Library of Liberal Arts, 211.)

105 "great secretary of nature": Henry More, quoted from J.L. Heilbron, "Was there a Scientific Revolution?" In Jed Buchwald and Robert Fox, eds., *The Oxford handbook of the history of physics*. Oxford: Oxford University Press, 2013, p. 16.

105 "oddest...detection": Newton to Oldenburg, January 18, 1671–2, in Isaac Newton, *The correspondence of Isaac Newton*. Ed. H.W. Turnbull et al. 7 vols. Cambridge: Cambridge University Press, 1959–77, 1, pp. 82–3.

110–11 "from the counsel," "I frame no hypothesis," and "[T]o us it is enough": Isaac Newton, *Mathematical principles of natural philosophy*. Tr. Andrew Motte (1729), revised Florian Cajori. Berkeley: University of California Press, 1934, pp. 544–7.

116 "in so far as there is mathematics in it": Immanuel Kant, *Philosophy of material nature*. Tr. James W. Ellington. Indianapolis: Hackett 1985, p. 6.

119–20 "a person would stand": Joseph Priestley, *The history and present state of electricity*. 2 vols. London: C. Bathurst et al., 1775, 2, p. 30.

122–3 "in every special doctrine" and "Chemistry can become nothing": Immanuel Kant, *Metaphysical foundations of modern science*. Tr. James W. Ellington. Indianapolis: Bobbs-Merrill, 1970, pp. 6, 7.

125 "thermometer of public prosperity": Karin Johannison, "Society in numbers: the debate over quantification in 18th-century political economy." In Tore Frängsmyr, J. L. Heilbron, and Robin E. Rider, eds., *The quantifying spirit in the 18th century*. Berkeley: University of California Press, p. 358, quoting Anon., *Recherches... sur la population de la France* (1778).

125 "Experiment, research": J.L. Heilbron, "The measure of Enlightenment." In Frängsmyr, Heilbron, and Rider, eds., *The quantifying spirit in the 18th century*, pp. 10–11, quoting Lavoisier.

CHAPTER 5

136 Poincaré, Henri. "Relations entre la physique expérimentale et la physique mathématique." In Congrès international de physique réuni à Paris en 1900. *Rapports et travaux*. 4 vols. Paris: Gautier-Villars, 1900–1, 1, pp. 1–29.

140 Kelvin, William Thomson, Baron. "Nineteenth century clouds over the dynamical theory of heat and light" (1900). In *Baltimore lectures on molecular dynamics and the wave theory of light*. London: C.J. Clay, 1904, pp. 486–527.

155 "who, during the preceding year": Nobel Foundation. *Directory*. Stockholm: Nobel Foundation, annual, p. 5.

155 "Science has reached a point": "Denkschrift...an den Kaiser," in Kaiser-Wilhelm-Gesellschaft, *50 Jahre Kaiser-Wilhelm-Gesellschaft und Max-Planck-Gesellschaft: Beiträge und Dokumente*. Göttingen: Max-Planck-Gesellschaft, 1961, p. 90.

CHAPTER 6

167 "'the old one'...is not playing at dice": Einstein to Born, December 4, 1926, in Albert Einstein and Max Born, *The Born–Einstein letters*. Tr. Irene Born. New York: Walker, 1971, p. 91.

173 "the elements originated": quoted in Helge Kragh, *Cosmology and controversy: the historical development of two theories of the universe*. Princeton: Princeton University Press, 1996, p. 105.

183 "In the public mind": Dyson, quoted in Silvan S. Schweber, *QED and the men who made it: Dyson, Feynman, Schwinger, and Tomonaga*. Princeton: Princeton University Press, 1994, p. 201.

CHAPTER 7

196 "tranquilizing philosophy": Einstein to Schrödinger, May 31, 1928, in Einstein, Albert, et al., *Letters on wave mechanics*. New York: Philosophical Library, 1967, p. 31.

198 "The more we know" and "The effort to understand": Steven Weinberg, *The first three minutes: a modern view of the origin of the universe*. 2nd edn. New York: Basic Books, 1993, pp. 154–5.

198-9   "can we actually say" and "we really do not understand": David Gross, "The major unknowns in particle physics and cosmology." In R.Y. Chiao et al., eds., *Visions of discovery: new light on physics, cosmology, and consciousness.* Cambridge: Cambridge University Press, 2011, pp. 153, 165.

FURTHER READING

GENERAL

Bearman, Peri J. et al., eds. *The encyclopedia of Islam*. Leyden: Brill, 2000– . Multivolume electronic resource.

Boyer, Carl B. *The rainbow: from myth to mathematics*. Basingstoke: Macmillan, 1987.

Buchwald, Jed, and Robert Fox, eds. *The Oxford handbook of the history of physics*. Oxford: Oxford University Press, 2013.

Bush, Steven. *A history of modern planetary physics*. 3 vols. Cambridge: Cambridge University Press, 1996.

Darrigol, Olivier. *A history of optics from Greek antiquity to the 19th century*. Oxford: Oxford University Press, 2012.

Dijksterhuis, E.J. *The mechanization of the world picture*. Oxford: Oxford University Press, 1961.

Frisinger, H. Howard. *The history of meteorology to 1800*. New York: Science History Publications, 1977.

Gillispie, C.C. et al., eds. *Dictionary of scientific biography*. 18 vols. New York: Scribners, 1970–81; continued by Noretta Koertge, ed., *New DSB*, 8 vols., 2008.

Heilbron, J.L., ed. *The Oxford companion to the history of modern science*. Oxford: Oxford University Press, 2003.

Kragh, Helge. *Conceptions of cosmos: from myths to the accelerating universe*. New York: Oxford University Press, 2007.

Kragh, Helge. *Higher speculations: grand theories and failed revolutions in physics and cosmology*. Oxford: Oxford University Press, 2011.

Lindberg, D.C., and R.L. Numbers et al., eds. *The Cambridge history of science*. 6 vols. (of 8). Cambridge: Cambridge University Press, 2003– .

North, John. *Cosmos: an illustrated history of astronomy and cosmology*. Chicago: University of Chicago Press, 2008.

Petruccioli, Sandro et al., eds. *Storia della scienza*. 10 vols. Rome: Istituto della Enciclopedia Italiana, 2001–4.

Rossi, Paolo et al., eds. *Storia della scienza moderna e contemporanea*. 6 vols. Milan: TEA, 2000.

CHAPTER 1

Cohen, Morris R., and I.E. Drabkin, eds. *A source book in Greek science*. Cambridge: Harvard University Press, 1958.

Lucretius. *On the nature of things*. Tr. Ronald Melville. Oxford: Oxford University Press, 2008.

Ptolemy, *Tetrabiblos*. Tr. F.E. Robbins. Cambridge: Harvard University Press, 1964.

Stahl, William H. *Roman science: origins, development and influence to the later Middle Ages*. Madison: University of Wisconsin Press, 1962.

Taub, Liba. *Ancient meteorology*. London: Routledge, 2003.

CHAPTER 2

Bloom, Jonathan M. *Paper before print: the history and impact of paper in the Islamic world*. New Haven: Yale University Press, 2001.

Ihsanoglu, Ekmeleddin. *Science, technology and learning in the Ottoman Empire*. Aldershot: Ashgate, 2004.

Makdisi, George. *The rise of colleges: institutions of learning in Islam and the West*. Edinburgh: Edinburgh University Press, 1981.

Makdisi, George. *The rise of humanism in classical Islam and the Christian West: with special reference to scholasticism*. Edinburgh: Edinburgh University Press 1990.

Masood, Ehsan. *Science and Islam: a history*. London: Icon, 2009.

Nasr, Sayyid Hussein. *Islamic science: an illustrated survey*. World of Islam, 1976.

Peters, F.E. *Aristotle and the Arabs*. New York: New York University Press, 1968.

Rashed, Rashdi, ed. *Histoire des sciences arabes*. 3 vols. Paris: Seuil, 1997.

Saliba, George. *Islamic science and the making of the European Renaissance*. Cambridge: MIT Press, 2007.

Sayili, Aydin. *The observatory in Islam*. New York: Arno, 1981.

Turner, Howard R. *Science in medieval Islam: an illustrated introduction*. Austin: University of Texas Press, 1997.

CHAPTER 3

Clagett, Marshall. *The science of mechanics in the Middle Ages*. Madison: University of Wisconsin Press, 1959.

Copernicus, Nicolas. *On the revolutions*. Tr. Edward Rosen. Baltimore: Johns Hopkins University Press, 1978.

Dante Alighieri. *The divine comedy*. Ed. Charles Singleton. 6 vols. Princeton: Princeton University Press, 1970–5.

Maier, Annaliese. *Studien zur Naturphilosophie der Spätscholastik*. 5 vols. Rome: Storia e letteratura, 1949–58.

Yates, Frances A. *Giordano Bruno and the Hermetic tradition*. Chicago: University of Chicago Press, 1964.

CHAPTER 4

Frängsmyr, Tore, J.L. Heilbron, and Robin E. Rider, eds. *The quantifying spirit in the 18th century*. Berkeley: University of California Press, 1990.

Friedman, Michael. *Kant and the exact sciences*. Cambridge: Harvard University Press, 1992.

Heilbron, J.L. *Electricity in the 17th and 18th centuries: a study in early modern science*. Berkeley: University of California Press, 1979.

Taton, René, ed. *Enseignement et diffusion des sciences en France au xviiie siècle*. Paris: Hermann, 1964.

CHAPTER 5

Buchwald, Jed. *From Maxwell to microphysics: aspects of electromagnetic theory in the last quarter of the 19th century*. Chicago: University of Chicago Press, 1985.

Buchwald, Jed. *The rise of the wave theory of light: optical theory and experiment in the early 19th century*. Chicago: University of Chicago Press, 1989.

Congrès international de physique réuni à Paris en 1900. *Rapports et travaux*. 4 vols. Paris: Gautier-Villars, 1900–1.

Darrigol, Olivier. *Electrodynamics from Ampère to Einstein.* Oxford: Oxford University Press, 2000.

Forman, Paul, J.L. Heilbron, and Spencer Weart. *Physics circa 1900.* Princeton: Princeton University Press, 1975. (Historical Studies in the Physical Sciences, vol. 5.)

CHAPTER 6

Dahl, Per F. *Superconductivity: its historical roots and development from mercury to the ceramic oxides.* New York: American Institute of Physics, 1992.

Fleming, James Rodger, ed. *Historical essays on meteorology, 1919–1995.* Boston: American Meteorological Society, 1996.

Forman, Paul. *National military establishments and the advancement of science and technology: studies in 20th century history.* Ed. J.M. Sanchez Ron. Dordrecht: Kluwer, 1996.

Forman, Paul. *Weimar culture and quantum mechanics: selected papers.* Ed. Cathryn Carson et al. London: Imperial College Press, 2011.

Glen, William. *The road to Jaramillo: critical years of the revolution in earth science.* Stanford: Stanford University Press, 1982.

Heilbron, J.L., and Robert W. Seidel. *Lawrence and his laboratory.* Berkeley: University of California Press, 1989.

Krige, John. *American hegemony and the postwar reconstruction of science in Europe.* Cambridge: MIT Press, 2006.

Krige, John, and Dominique Pestre, eds. *Science in the twentieth century.* Amsterdam: Harwood, 1997.

Mendelssohn, Kurt. *The quest for absolute zero.* London: Taylor & Francis, 1977.

INDEX

GALILEO

John L. Heilbron

978-0-19-965598-4 | Paperback | £12.99

'Heilbron has produced that rare marvel, a splendid new view of a familiar figure, a witty, absorbing, and convincing account of the man and his epoch, destined for the wide readership Galileo himself once had.'

Eileen Reeves, *ISIS*

Galileo's *Siderius nuncius*, or Starry Messenger, presented to the world his remarkable observations using the recently invented telescope—of the craters of the moon, and the satellites of Jupiter, observations that forced changes to perceptions of the perfection of the heavens and the centrality of the Earth. Yet he was no more (or less) a mathematician than he was a musician, artist, writer, philosopher, and gadgeteer.

This fresh lively biography of the 'father of science' paints a rounded picture of Galileo, and places him firmly within the rich texture of late Renaissance Florence, Pisa, and Padua, amid debates on the merits of Ariosto and Tasso, and the geometry of Dante's Inferno.

THE INFINITY PUZZLE

*The personalities, politics, and extraordinary
science behind the Higgs boson*

Frank Close

978-0-19-967330-8 | Paperback | £10.99

'Thoroughly researched and well-crafted narrative.'
New Scientist

'Masterpiece ... I never normally give 5 stars but for this I make an exception.'
John Gribbin, *BBC Focus*

Forty or so years ago, three brilliant, yet little-known scientists—an American, a Dutchman, and an Englishman—made breakthroughs which later inspired the construction of the Large Hadron Collider at CERN in Geneva: a 27 kilometre-long machine that cost ten billion dollars, took twenty years to build, and finally discovered a particle consistent with the Higgs boson. Played out across the decades against a backdrop of high politics, low behaviour, and billion dollar budgets, *The Infinity Puzzle* draws upon the author's close friendships with those involved.

Sign up to our quarterly e-newsletter **http://academic-preferences.oup.com/**

NEUTRINO

Frank Close

978-0-19-969599-7 | Paperback | £7.99

'Close tells this story with verve and precision...
admirably clear and eminently accessible.'

Wall Street Journal

'As an award-winning writer, Close tells this
detective story with great style.'

Robert Matthews, *BBC Focus*

Neutrinos are perhaps the most enigmatic
particles in the universe. Formed in certain
radioactive decays, they pass through most
matter with ease. These tiny, ghostly particles
are formed in millions in the Sun and pass
through us constantly. For a long time they
were thought to be massless, and passing as
they do like ghosts they were not regarded as
significant. Now we know they have a very
small mass, and there are strong indications that
they are very important indeed. It is speculated
that a heavy form of neutrino, that is both matter
and antimatter, may have shaped the balance of
matter and antimatter in the early universe.

Here, Frank Close gives an account of the
discovery of neutrinos and our growing understanding of their significance, and also touches on
some speculative ideas concerning the possible
uses of neutrinos and their role in the early
universe.

Sign up to our quarterly e-newsletter **http://academic-preferences.oup.com/**

ORIGINS

The scientific story of creation

Jim Baggott

978-0-19-870764-6 | Hardback | £25.00

'The collective mind of humanity has made extraordinary progress in its quest to understand how the current richness of the physical world has emerged, and Baggott with his characteristic lucidity and erudition, has provided an enthralling account of this wonderful and still unfolding intellectual journey.'

Peter Atkins

There are many different versions of our creation story. This book tells the version according to modern science', writes Jim Baggott. In *Origins*, he presents a unique version of the story in chronological sequence, from the Big Bang to the emergence of human consciousness 13.8 billion years later.

Cosmology, particle physics, chemistry, planetary geology, biology—it is all here, explained with clarity, in one overarching narrative. And throughout, Baggott emphasizes that the scientific story is a work-in-progress, highlighting the many puzzles and uncertainties that still remain. We have a seemingly innate desire to comprehend our own place in the Universe. Jim Baggott helps us fulfil this desire, which is driven in part by simple curiosity but also by a deeper emotional need to connect ourselves meaningfully with the world which we call home.

PLANCK

Driven by vision, broken by war

Brandon R. Brown

978-0-19-021947-5 | Hardback | £20.00

'An illuminating biography...Brown interweaves a gripping backstory, ranging from Planck's landmark theoretical description of blackbody radiation to his loyal advocacy for fellow physicist Lise Meitner.'

Nature

Nobel prizewinning German physicist Max Planck is credited with being the father of quantum theory, while Planck's Law was described by his close friend Albert Einstein as 'the basis of all twentieth-century physics'. But despite being one of the great scientists, Planck's personal story is not well known. Brandon R. Brown interweaves the voices and writings of Planck, his family, and his contemporaries—with many passages appearing in English for the first time—to create a portrait of a ground-breaking physicist working in Germany throughout two world wars. This gripping story of a brilliant man living through dangerous times gives Max Planck his rightful place in the history of science.

Sign up to our quarterly e-newsletter **http://academic-preferences.oup.com/**

THE END OF DISCOVERY

Are we approaching the boundaries of the knowable?

Russell Stannard

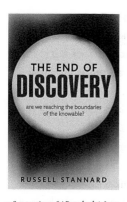

978-0-19-964571-8 | Paperback | £9.99

'Lucid and provocative, it is a very polite corrective to both superstitions of the layman and the triumphalism of the experts.'

New Statesman

It is generally thought that science, by its very nature, must always progress. But this is not so. One day, fundamental science will come to an end. Not when we have discovered everything, but when we have discovered whatever is open to us to understand—which is not the same thing.

Author and broadcaster Russell Stannard introduces the deepest questions facing us today—questions to do with consciousness, free will, the nature of space, time, and matter, the existence of extraterrestrial life, and why there should be a world at all. In doing so, he speculates as to whether some of these questions will never be answered.

Sign up to our quarterly e-newsletter **http://academic-preferences.oup.com/**

THE QUANTUM STORY

A history in 40 moments

Jim Baggott

978-0-19-965597-7 | Paperback | £14.99

'Jim Baggott's survey of the history of the emergence of the twentieth century's most enigmatic but successful theory is a delight to read. It is clear, accessible, engaging, informative, and thorough. It illuminates an important, revolutionary era of modern science and the varied personalities behind it.'

Peter Atkins

Almost everything we think we know about the nature of our world comes from one theory of physics. Jim Baggott presents a celebration of this wonderful yet wholly disconcerting theory, with a history told in forty episodes—significant moments of truth or turning points in the theory's development. From its birth in the porcelain furnaces used to study black body radiation in 1900, to the promise of stimulating new quantum phenomena to be revealed by CERN's Large Hadron Collider over a hundred years later, this is the extraordinary story of the quantum world.

Sign up to our quarterly e-newsletter **http://academic-preferences.oup.com/**